Design of Experiments in Quality Engineering

Jeffrey T. Luftig, Ph.D.

Victoria S. Jordan

McGraw-Hill

New York San Francisco Washington, D.C. Auckland Bogotá
Caracas Lisbon London Madrid Mexico City Milan
Montreal New Delhi San Juan Singapore
Sydney Tokyo Toronto

Library of Congress Cataloging-in-Publication Data

Luftig, Jeffrey T.
  Design of experiments in quality engineering / Jeffrey T. Luftig, Victoria S. Jordan.
       p.   cm.
  ISBN 0-07-038807-5
  1. Research, Industrial.   2. Experimental design.   I. Jordan, Victoria S.   II. Title.
  T175.L83   1998
  607'.24—dc21   97-24465
                                                                   CIP

McGraw-Hill

A Division of The McGraw-Hill Companies

Copyright © 1998 by The McGraw-Hill Companies, Inc. All rights reserved. Printed in the United States of America. Except as permitted under the United States Copyright Act of 1976, no part of this publication may be reproduced or distributed in any form or by any means, or stored in a data base or retrieval system, without the prior written permission of the publisher.

1 2 3 4 5 6 7 8 9 0   DOC/DOC   9 0 2 1 0 9 8 7

ISBN 0-07-038807-5

The sponsoring editor for this book was Harold B. Crawford, the editing supervisor was Penny Linskey, and the production supervisor was Pamela Pelton. It was set in Century Schoolbook by Teresa F. Leaden of McGraw-Hill's Professional Book Group composition unit.

Printed and bound by Donnelley/Crawfordsville.

McGraw-Hill books are available at special quantity discounts to use as premiums and sales promotions, or for use in corporate training programs. For more information, please write to the Director of Special Sales, McGraw-Hill, 11 West 19th Street, New York, NY 10011. Or contact your local bookstore.

This book is printed on recycled, acid-free paper containing a minimum of 50% recycled, de-inked fiber.

Information contained in this work has been obtained by The McGraw-Hill Companies, Inc. ("McGraw-Hill") from sources believed to be reliable. However, neither McGraw-Hill nor its authors guarantees the accuracy or completeness of any information published herein and neither McGraw-Hill nor its authors shall be responsible for any errors, omissions, or damages arising out of use of this information. This work is published with the understanding that McGraw-Hill and its authors are supplying information, but are not attempting to render engineering or other professional services. If such services are required, the assistance of an appropriate professional should be sought.

Contents

Acknowledgments   vii

Introduction   1

Chapter 1. Types and Purposes of the Research Study   7

Describing the Research Study   7
   The role of experimental design in industrial research   7
   The scientific basis for industrial research   9
   Defining the research problem   14
   Developing a statement of the problem   17
Developing the Theoretical Framework of the Research Study   20
   Analytical research   25
   Agreement research   26
   Descriptive research   27
   Relational research   28
   Experimental research   32
Making the Statement of the Problem Operational: Research Questions and Hypotheses   32
Criteria for Evaluating Research Questions and Hypotheses   34
Checklist for Planning Industrial Research Studies   36

Chapter 2. Developing the Experimental Design: Concepts and Definitions   39

Underlying Concepts   39
Basic Definitions of Terms   40
Supporting Definitions   45

Chapter 3. An Introduction to Statistical Inference   47

Inferential Statistical Methods   47
The One-Way Analysis of Variance (ANOVA)   57

Chapter 4. Developing the Experimental Design: Defending the Design Against Internal and External Threats to Validity   61

The Basic Logic and Purpose of an Experimental Design 61
Type I error: The treatment is not related to/does not affect the dependent variable, but the experiment causes the researcher to infer that a relationship/effect does exist 62
An overview of the threats to internal validity 69
Designs which protect against major threats to internal validity 73
Type II error: The treatment is related to/does affect the dependent variable, but the experiment causes the researcher to infer that no relationship/no effect exists 76
Power: The treatment is related to/does affect the dependent variable, and the experiment results in the identification of the effect 95
Confidence: The treatment is not related to/does not affect the dependent variable, and the experiment results in this conclusion 96
Threats to External Validity 97
Types of Experimental Designs 98

Chapter 5. Steps for Designing and Assessing an Industrial Experiment 105

Appropriate Steps for Designing an Industrial Experiment 105
Step 3: Define the dependent variables and their associated criterion measures 106
Step 4: Identify and classify the independent variables 107
Step 5: Select the levels associated with each treatment variable 118
Step 6: Select the experimental design appropriate to the treatment variables and number of levels selected; modified as required, and assign the treatments to the design plan 120
Assessing the Industrial Experiment for Adequacy and Efficiency 122
Updating the Planning Checklist for Industrial Research 124

Chapter 6. Sampling Procedures and Considerations 131

Types of Sampling Plans and Methods 131
Nonprobabilistic sampling plans 131
Probabilistic sampling plans 133
Misconceptions Related to the Topic of Randomness 135
Sample Size and the Precision of the Experiment 136
Summary 139

Chapter 7. Establishing the Validity of the Data 141

Categories of Data 141
Effectiveness of the Instrumentation 142
Reliability 143
Validity 144
Updating the Planning Checklist for Industrial Research 145

Chapter 8. Managing the Execution of the Experiment 155

Considerations and Responsibilities 155
The Engineering Log 156

Chapter 9. Designing the Plan for the Statistical Analysis of the Data 161

Measurement Scales and Test Selection 161

Parametric Testing and Analysis — 162
Nonparametric Testing and Analysis — 163
The Seven-Step Procedure for Hypothesis Testing — 163
    Step I: State the null (H_o) and research (H_1) hypotheses — 167
    Step II: State the maximum risk of committing a Type 1 error previously selected — 167
    Step III: State the associated test statistic — 168
    Step IV: Identify the random sampling distribution (RSD) of the test statistic when H_o is true — 168
    Step V: State the critical value for rejecting the null hypothesis — 168
    Step VI: Calculate the value of the test statistic from the sample data — 169
    Step VII: Analyze the results and make an appropriate decision related to the null hypothesis — 169

Chapter 10. Reporting and Standardizing the Results of the Research Study — 171

Reporting the Results of the Study — 171
The Standardize–Do–Check–Act (SDCA) Process — 173
The Final Checklist for the Planning of Industrial Research — 173

Chapter 11. Designing the Industrial Experiment: Case Studies — 189

Case Study 1: A Completely Randomized Design — 190
Case Study 2: The Randomized Block Design: Matched Pairs — 201
Case Study 3: A Randomized Block Design for More Than Two Treatment Levels — 210
Case Study 4: A Complete Block Design for More Than Two Blocked Effects — 215
    Latin square design — 216
    Youden square designs and other variations of the Latin square design — 221
Case Study 5: Factorial Experiments—A fully Crossed Type I Model — 221
Case Study 6 and Self-Review Activity 11.7: Factorial Experiments—A Fully Crossed Type I Model — 229
Case Study 7: Factorial Experiments—A Model III Nested Design — 240
Case Study 8: Fractional Factorial Experiments — 253
    Extreme screening designs — 256
    Higher-resolution fractional factorial designs — 257
Additional Experimental Design Approaches — 263
    Analysis of covariance — 264
    Evolutionary operation (EVOP) technique — 265

Appendix A: A Planning Checklist for Industrial Research — 267

Appendix B: Sources of Invalidity for Some Preexperimental, Experimental, and Quasi-Experimental Designs — 281

Appendix C: Selected Tables of Critical Values — 285

Table C-1. Table of areas under the normal curve — 286

Table C-2. Percentage points, student's t distribution (upper-tail probabilities) — 289

Table C-3. Percentage points. F distribution. $\alpha = 0.10$ (upper-tail) — 290

Appendix D: Blank Forms for the Seven-Step Procedure for Hypothesis Testing 299

Appendix E: Answers to Self-Review Activities 305

Bibliography 333
Index 335

Acknowledgments

The authors gratefully acknowledge the contributions of Dr. Jerry Streichler for his work as a consulting editor, Diane Copty-Luftig for her work in editing the final manuscript, and the administrative staff of Luftig & Warren International for their assistance with graphics, tables, figures, and reference information.

Introduction

The need for *Design of Experiments in Quality Engineering* can best be demonstrated by an anecdote:

> A man working in the production area set up an experiment to test a new alloy, to possibly replace one currently in production. He ran one heat of metal with the new alloy and another heat with the old one. Taking one ingot from each heat and 30 pieces of metal from each ingot, he proceeded to test each of the 60 pieces for the property in which he was interested. With the data he ran a statistical analysis [a one-way analysis of variance (ANOVA)] on the alloys using the pieces of ingots with 58 degrees of freedom as the error. The results showed that the new alloy was "better" than the old one. The experimenter convinced the vice president in charge of production to change the production procedures so that the new alloy would be used in the future. Since the experimenter used a "designed experiment" and tested the data "statistically," the vice president concluded that there could be no doubt that the new one was better.
>
> This change cost the company $200,000. After 2 years in the field there was as much trouble with the product made from the new alloy as there had been with the old product. The vice president was disgusted and declared that he would never allow his company to use designed experiments again (Anderson and McLean, 1974).

Despite the fact that this was written more than twenty years ago, most individuals working in industry today would not find this anecdote surprising or unique. In fact, the lack of rigor and discipline displayed in many companies in the conduct of industrial research is astonishing. Slipshod techniques and methods, inappropriate and incomplete statistical analyses, and unwarranted conclusions and assertions are common. It seems that any conscientious manager would ask: "Why? Why are such conditions so prevalent?"

The answer to this question may be traced to a number of elements:

1. There is a general lack of formal education on the part of managers and supervisors in the area of industrial research methods.
2. Engineers and other technical personnel trained in statistical methods often have little education in experimental design methods, and even less in research design technology.
3. Outside of research and development departments and sales and marketing groups, the use of research and experimental methods on a day-to-day basis for data-based decision making is simply not common in the management process.

This book provides managers and supervisors with a much needed understanding and appreciation of the industrial research process. Further, it supplies the reader with a series of "checklists" that correspond to each step of the research process. This knowledge will allow readers to more appropriately direct and participate in industrial research in their areas of responsibility. Although statistics are used to describe procedures, this book does not concentrate on the field of industrial statistics. Rather, it stresses the appropriate approach to industrial research and experimentation.

The material for this book is based on a successful marriage of two fields: (1) research and statistics and (2) interfacing engineering and industrial technology with management. Over the past 20 years, the authors have applied knowledge and skills of these fields in service to a number of major domestic and international companies.

This book shares some of the results of that experience. It has also been used successfully in training programs in which practicing managers, supervisors, and technical personnel have increased their awareness and improved their capabilities. Knowledge and skills that they have gained in the area of industrial experiments have allowed them to better fulfill responsibilities for diagnosing and solving process and product problems and for objectively assessing the relative qualities and capabilities of materials and equipment offered by competing vendors. As a result, they have become better engineers, technicians, and managers and have clearly become better able to contribute to the effectiveness and profitability of their respective organizations.

In this book, engineering and management decision makers are provided with knowledge that will enhance capabilities in design of experiments. Readers will develop a substantial understanding, far beyond awareness, that will enable and facilitate decisions concerning when and where design of experiments should be applied and whether the process is applied correctly and effectively. It should also allow the decision maker to select capable personnel who will administer design of experiments within the organization's quality effort, or prescribe training to produce such people.

The Approach

The reader is transported in actual cases from the basics of industrial research to applications of sophisticated techniques. The cases provide insight into the varieties of techniques, processes, and treatments. They also show how cogent selections are made by the researcher to ensure that the design of the research is indeed focused on the knowledge to be gained and/or problems to be solved. The cases further demonstrate the important role of research in decision making to quality, productivity, and profitability.

The relationships and the techniques of design of experiments are also considered as they are manifested within quality function deployment and statistical process control so as to depict the entire spectrum of the subject in all the components of an organization's statistical quality control effort. Design of experiments, when properly implemented, and the industrial research technique presented in this book are applicable to virtually any preproduction, production, and postproduction activity, such as product and process improvement, equipment selection, product design, vendor/product analysis, maintenance techniques, equipment selection/effectiveness, customer comparisons, and cost-effectiveness.

Contents

The following is a summary of each chapter.

Chapter 1. Types and purposes of the research study. This chapter provides background on industrial research and outlines the process for defining the research hypothesis and statement of the problem (steps 1 and 2 for designing an industrial experiment).

Chapter 2. Developing the experimental design; the concept and associated definitions. This chapter provides definitions of terms used in this text that are applicable in the field of industrial research.

Chapter 3. An introduction to statistical inference. This chapter introduces the reader to the concept of statistical interference and the one-way analysis of variance.

Chapter 4. Developing the experimental design; the purpose and types of designs. Chapter 4 provides an overview of the different types of designs and how they can be used to avoid errors in industrial research. Threats to internal and external validity are identified and addressed with the proper experiment design.

Chapter 5. Steps for designing an industrial experiment. Appropriate steps for designing an industrial experiment are identified. Defining the research hypothesis and statement of the problem (steps 1 and 2) were studied in Chap. 1. This chapter covers the remaining steps: step 3: define the dependent variables and the associated criterion measures; step 4: identify and classify the independent variables; step 5: select the levels associated with each treatment variable; and step 6: select the experiment design appropriate to the treatment variables and number of levels selected, modify as required, and assign the treatments to the design plan. The industrial experiment is also assessed for adequacy and efficiency.

Chapter 6. Sampling procedures and considerations. Types of sampling plans (nonprobabilistic and probabilistic plans) and methods are defined.

Misconceptions related to the topic of randomness, sample size calculations, and the precision of the experiment are explored as they affect the validity of the design.

Chapter 7. Establishing the validity of the data. This chapter identifies the categories of data and the effectiveness of the instrumentation, and the effect they have on the experiment design.

Chapter 8. Managing the execution of the experiment. Once the experiment is designed, it must be properly executed to yield valid results. This chapter identifies considerations and responsibilities in the execution of the experiment and introduces the engineering log as an important method for monitoring variables during the experiment.

Chapter 9. Designing the plan for the statistical analysis of the data. This chapter explains how to select an appropriate statistical analysis and how to use the seven-step procedure for hypothesis testing.

Chapter 10. Reporting and standardizing the results of the research study. This chapter shows how the information should be reported after the experiment is concluded and the results are statistically analyzed. The standardize–do–check–act (SDCA) process is introduced to standardize any improvements resulting from the experiment.

Chapter 11. Designing the industrial experiment: case studies. Case studies are presented for various types of experiment design including a completely randomized design, a randomized block design—matched pairs, a randomized block design for more than two treatment levels, a complete block design for more than two blocked effects (the Latin square), factorial experiments—a Model III nested design, and fractional factorial experiments—extreme screening designs, higher resolution designs (IV and V). Analysis of covariance designs and evolutionary operations research are also presented.

Appendices. The appendices include a planning checklist for industrial research, sources of invalidity, blank seven-step procedure forms, statistical tables, and answers to the self-review activities that are presented throughout the book.

Using *Design of Experiments in Quality Engineering*

The design of this book allows readers to achieve goals based on their interests, responsibilities, and aspirations. It can be read (and used) to acquire a better general knowledge of the role and uses of design of experiments in industry. It will allow the reader who seeks to develop skills to do so and to develop a keener understanding of how to implement a program of industrial research that will contribute to the achievement of an organization's quality and profitability goals. In short, it will benefit a wide range of individuals with managerial or technical responsibilities. Since the end use of the skill and knowledge that will be acquired from reading this book is likely to be different for different readers, reading and using this book may be approached differently to achieve the different goals. Thus:

1. The roles and functions of research in general are described and defined. This provides the context for industrial research that is generally described and then in turn establishes the context for industrial experiments, which is treated in greater detail.
2. Just by reading, it is possible to progress through the book. However, more can be gained by using the opportunities to interact with the material that is structured into the volume. This can be approached in three ways:
    a. Self-review activities are provided at key points in the text. These are structured to challenge the reader to apply what has been learned. The questions, problems, and activities are based on the material that has been covered. But more important, whenever possible, they cause the reader to do and act on real and meaningful problems because they are all based on actual issues and challenges encountered in industry.
    b. As the reader progresses through the text, a checklist for planning industrial research provides another opportunity to interact and apply the knowledge gained from the text material. The checklist parallels and grows with the content. Thus, the first one that appears in the text is repeated and added to in those checklists that follow until a comprehensive document results at the end of the text material.
    c. The cases that are included are based on actual situations. They are arranged to exemplify different problems and issues that require different research designs and treatments. Interaction between the text material and the material presented in the cases is encouraged in two ways. First, the cases are presented in such a way that at crucial times reference is made to the relevant text material. Second, some cases include self-review activities that are structured to motivate the reader to relate the material of the case to the relevant material of the text.

Experiment Design in Quality Engineering provides engineering and management decision makers with the knowledge and tools to determine when industrial research, specifically design of experiments, is appropriate and how to apply the research process correctly.

Chapter

1

Types and Purposes of the Research Study

Describing the Research Study

The role of experimental design in industrial research

When we speak of research, we refer to any activity that produces knowledge. As translated from its Latin roots, the term *research* implies the "search for the truth." The two generally recognized domains of research are basic and applied. Basic research is designed to develop theoretical knowledge. It is usually not intended or designed to solve problems or answer any specific questions. Applied research, on the other hand, is designed to answer specific questions or test particular hypotheses that have immediate application to practice (Sax, 1979).

A number of types of research are discussed in detail later in this book. Although this book concentrates on experiments, we note that an experiment is only one type of research. When we use the term *research design,* we refer to a detailed methodology or plan of study used in conducting research of any type. An *experimental design* describes a plan of narrower context (Miller, 1964). It refers to a plan of study determined by a statistical procedure. Kirk (1968) tells us that experimental design designates a type of plan to be used for the assignment of test units (or subjects) to experimental conditions (or treatments) for the purpose of generating (versus collecting) data. He also considers the sampling plan, which deals with numbers of procedures or processes that will receive some sort of "treatment," and the statistical analysis plan to be part of the experimental design. While these serve to expand the term somewhat, experimental design still describes an activity of a more limited nature than that implied by the term research design.

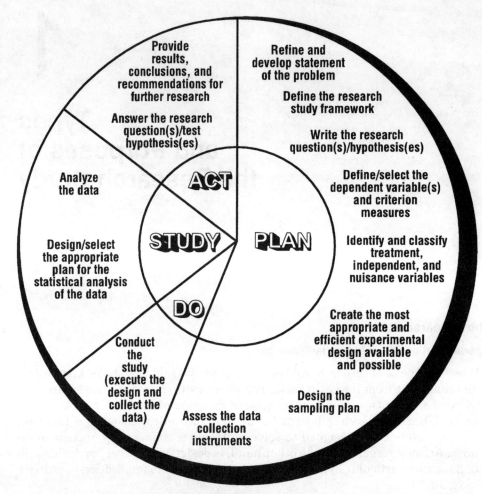

Figure 1.1 Overview of the Research Design Process.

Figure 1.1 further defines the context and interrelationship of these terms. It shows that the research design process may be effectively organized in terms of the plan–do–study–act (PDSA) management discipline. This figure also provides an overview of the major steps in the research process, specifically as associated with the conduct of experiments.

As you review this illustration, note the following characteristics in the research process:

1. The proportion of the process associated with the planning of the study.
2. The role, size, and placement of experimental design within research design.
3. The distance within the research design that the collection and statistical analysis of the data appears. This location, symbolically more than halfway

into the process, has prompted the caution that "Any study that begins with the collection of data is inevitably doomed to failure."

The scientific basis for industrial research

Whether conducted in industry, the social and behavioral sciences, or in services such as education, research is a process dedicated to the search for knowledge. While most of us may have some understanding of the research process as a disciplined, scientifically based effort, this view has not always been the norm. In 1897, for example, a government official named J.M. Rice posed the following question to leaders in the field of education and psychology

> "...how would it be possible to determine if students who are given 10 minutes of spelling each day learn any less than those who are given 40 minutes each day?"

This query, of course, is conceptually no different than asking

> "How would it be possible to determine whether product manufactured with incoming material from Vendor A will be of higher quality than product manufactured with incoming material from Vendor B?"
> or
> "How would it be possible to determine whether dome strength varies for tooling that is chrome plated versus tooling that is not chrome plated?"

When Rice posed his question at the end of the nineteenth century, however, many individuals responded that this question was probably not one that could be easily or conclusively answered. In fact, it was reported that one well-known professor of psychology indicated that such a question could never be answered, that it was a waste of time to even consider the issue (Rice, 1913).

At this point in time, the prevailing attitude was that there were four sources of knowledge that one could turn to for answers to research questions or problem-solving. Sax (1979) describes these sources as follows.

The appeal to common sense. As a source of knowledge, the appeal to common sense may be subdivided into two categories. The first would relate to knowledge obtained through experience or a body of previously investigated beliefs. As new information is obtained, the body of knowledge is modified. The second category would relate to those principles or beliefs that today may be preceded with the phrase "everyone knows that...." Unfortunately, this second category often prevents the appropriate study of those assumptions and truths that are widely accepted, but are, in fact, false.

The appeal to authority. As a source of knowledge, the appeal to authority is in wide use today in virtually every field of endeavor, and may be just as effectively abused as beneficially used. It is absolutely true that, in many cases, this may constitute an invaluable source of knowledge. It is also true that throughout history the declarations of those in authority have often been employed to quell inquiry and suppress dissent. One of the interesting aspects of this source of knowledge (versus power) is the determination of whether any individual should or should not be considered an authority in his/her field of endeavor or expertise. In the book *Logic for the Millions,* Mander (1947) suggested four criteria for the assessment of whether an individual should be considered an authority:

1. The authority should be an identifiable person (as opposed to "many specialists believe that…").
2. The authority should be recognized as such by members of the same profession or field in which extraordinary competency is claimed by the individual in question.
3. The authority should be living. While this criterion may be viewed in a relative context, the intent is clear. Madame Curie may have been considered an authority on radiation in her own time, but her knowledge and work would not likely be considered to be representative of exceptional expertise in this field today.
4. The authority should not be biased. Not to be confused with possessing a philosophical orientation or belief structure, this criterion would probably be the most difficult to assess.

The appeal to intuition and revelation. Sax (1979) observes that knowledge resulting from "direct and immediate insights concerning truth…presumed to originate from God." Revelations from natural origins are termed *intuitions.* Intuitive knowledge should rarely be accepted without empirical testing. Truth based on revelation is from some source outside of the human experience, and therefore not generally subject to empirical analysis and testing.

The appeal to logic. The appeal to logic is based primarily on rationalism, which values reason (predominantly deductive) over both experience and intuition. We may define deductive reasoning as the development of specific conclusions or applications from an accepted principle or premise. The deductive process is perhaps best represented by a syllogism. One of the most frequently employed examples of a syllogism appears as:

A is always B (the major premise), and

if C is also A (the minor premise), then

C must also be B

Deductive reasoning, taken in isolation, is not a completely inappropriate method of determining truth or deriving knowledge. It cannot, however, be considered to be a completely satisfactory method of determining the truth in the absence of empirical analysis. Logic may yield premises and applications that are intellectually reasonable, yet bear little resemblance to reality, particularly if the major premise involved is flawed.

Consider the syllogism:

The earth is flat

Everyone we know sailing to the Far East has never returned, therefore

Investing in Columbus is foolhardy

Yet, it was the appeal to logic and deductive reasoning that provided a major building block for the utilization of the scientific method as a source of knowledge.

The scientific method is defined by different authors in a number of fields in varying ways. The consensus is unclear and we may conclude that there is no universally held definition. We define the scientific method as that process by which deductive and inductive reasoning methods are employed to empirically develop knowledge. Inductive reasoning refers to the generation of general principles, or major premises, from specific known facts or observations. This process might best be illustrated by the following observational sequence:

Fact 1:	Field failures on a particular product have suddenly and dramatically increased.
Fact 2:	The design of the product was not changed.
Fact 3:	The plant production processes are still in a state of control, and have not been modified.
Fact 4:	The product is manufactured with a compound that is formulated and processed by an outside vendor.
Fact 5:	The outside vendor has, in the past, been known to make changes in the formulation of the compound, without informing the plant.
Therefore,	
Major premise:	The vendor has once again altered the purchased compound, causing an increase in the field failure rate.

Note that, like deductive reasoning, the inductive method of generating knowledge may exhibit significant shortcomings if used in isolation. If all of the relative facts in a given case are not known, the principle or premise developed may be rational and logical, but incorrect. For example, in the above sequence, suppose that the vendor did not change the compound. Also, suppose that a mating part to the product had been changed, and no one was aware of the design change. In this event, the premise developed would be logical, but incorrect.

Researchers in various fields (predominantly agriculture and manufacturing) began to move away from rationalism and toward the use of the scientific method at the turn of the century. During this period, the scientific method combined deductive and inductive reasoning principles in the context of the trial and error method of generating knowledge.

The trial and error method of generating knowledge is generally considered to be too expensive and time-consuming to be utilized today. Yet, as a result of some of the conditions detailed in the Introduction of this book, much of the research presently conducted in American industry today is performed by utilizing this process. How many times have you heard someone suggest, for example, "Let's change it and see what happens?" The trial and error method is utilized when no prior knowledge of the situation exists, and when the investigator has virtually no idea where to begin the inquiry. From a random starting point, the research is conducted through a series of approximations, with the results of each trial forming a block in the base for the structure of the following iteration.

An example of this type of research would be the studies conducted in the early 1900s in the area of polymer technology, particularly in Europe. Research in this new field often consisted of randomly mixing various materials of random volumes together, heating them, and seeing what happened. Remarkably, the first urea formaldehyde polymer was developed in Germany during the early 1900s using this method of generating knowledge.

After the trial and error period of inquiry was a period of further development referred to as the *traditional period*. This line of research employed inductive and deductive reasoning and empirical methods in primarily a one-factor-at-a-time process of testing. This method is contrasted with the modern approach in Chap. 4.

The modern approach to the scientific method combines the deductive reasoning process with the inductive reasoning process, and adds the use of appropriate and advanced experimental design conditions to yield a comprehensive, complex, and highly disciplined method of inquiry. The research design process model presented in Fig. 1.1 is an example of the application of the modern scientific method to the research process.

In this model, the inductive reasoning process is used to gather facts and generate the statement of the problem as well as the hypotheses to be tested. The deductive reasoning process is also used to develop alternative experimental designs necessary to test the hypotheses in question, and to develop conclusions based on the data collected.

While no single universally accepted definition of the scientific method is available, there is general agreement on those elements or conditions upon which the scientific method is based. The *Encyclopedia of Statistical Sciences* (Kotz and Johnson, 1985) provides the generally accepted attributes for a research model based on the modern (versus traditional) scientific method. As you review structure and content in the model for the research design process, compare them with the following criteria from the encyclopedia:

1. The model requires careful and accurate observations, stressing the quest for precision as well as the reduction and control of bias.
2. The model may include elements of verification: the repetition of another researcher's experiences, results, and conclusions. Additionally, the model requires that observations be reported in such a way so that other researchers may confirm or reject the results.

3. The model includes the opportunity for the researcher to capture unplanned correlations (relationships) among observed phenomena. These observed correlations may lead to inductively derived conjectures of causal relationships that may be subsequently verified by additional study. Examples of this condition include the discovery of x-rays by Röntgen in 1895 and penicillin by Fleming in 1928. The use of an engineering log (discussed in Chap. 8) is the first step in preparing for such an outcome.
4. The model is based upon empiricism. This is the position that observation and testing are the primary and preferred methods of obtaining knowledge. Further, empiricism states that in the presence of a conflict, knowledge acquired through less rigorous and disciplined methods (e.g., intuition and common sense) is relegated to a lower priority than knowledge gained through empirical methods.
5. The model is often oriented toward elements such as explanation, control, and prediction, based on existing theory (profound knowledge).
6. The model favors the generation of new theories based upon the law or principle of parsimony. This premise states that a simple or elegant theory is preferred to a complex or ugly one if both explain the same facts to the same degree. This premise does not preclude the development of complex constructs; it simply favors the simple over the complex if both are equal in explanation.
7. The model generally includes the generation of deductions from theories or hypotheses where those deductions may be predictive or explanatory in nature.
8. The model generally includes an attempt to provide a basis for the confirmation or refutation of the generated deductions through the use of further experiments and observations. (Examine the ACT portion of the research process presented in Fig. 1.1.)
9. The model should stress the use of rationality versus rationalism in obtaining and analyzing the generated observations. In other words, wishful thinking is to be avoided, objectivity is to be stressed.
10. The model should include the effective communication of the results of the research. In academia, the incentive for this element is to "publish or perish." In industry, research is generally conducted for a specific purpose with profitability, productivity, or survival often hanging in the balance. As a result, any research model that is to be effective in industry must include this attribute.

This brings us to the interface of the research design process (as based on the modern scientific method) and the conduct of applied research in the industrial setting. We have discussed the nature of the research process as based on the scientific method. The purpose of the research process is to provide answers to research questions. This leads us to the need to establish the source of the research question, as expressed by the statement of the problem.

14   Chapter One

Defining the research problem

The major thrust in this book relates to applied, rather than basic, research. In this context, there are three major sources of research problems that would result in the design of a research study and the subsequent development of a statement of the problem.

1. Strategic options and initiatives generated through product-market analysis and/or a technical competitive benchmarking effort. These options and initiatives often yield product or process shortcomings, or gaps, that a company would frequently target for closure through applied research.

2. Strategic research questions generated while conducting quality improvement activity through the quality improvement strategy (QIS). Frequently, a quality improvement team (QIT) or natural work group (NWG) will use the QIS, depicted in Fig. 1.2, to bring a critical quality characteristic into a state of control and capability. In the course of their effort, they may find that an applied research activity will be required. Some of the anticipated

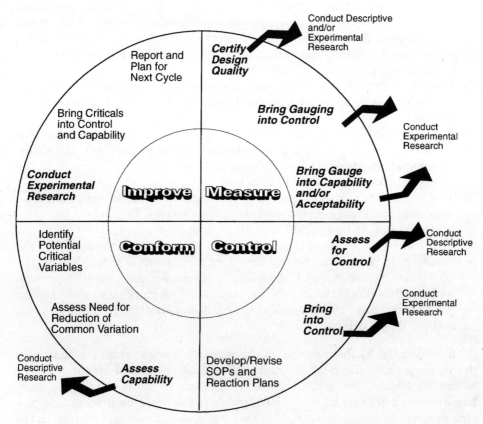

Figure 1.2 Research Design Applications in the Quality Improvement Strategy as Applied to Product Quality.

points at which this event may be expected to occur are also shown in Fig. 1.2. (The figure provides a breakdown of research applications into descriptive and experimental categories. These terms are fully defined later in this chapter.) For example, if the design quality* of a particular product has never been established, some type of research study might be initially employed to verify that the engineering targets are, in fact, optimal.

At this juncture of the QIS, there are many cases where a QIT might conduct a similar study as it attempts to find ways to modify quality characteristic targets so as to arrive at a more robust product design. This principle would also hold true for process design applications. For example, an NWG operating in a wave soldering department might use applied research methods to determine process set points that would allow them to operate in a state that was robust against uncontrollable shifts in ambient temperature and humidity.

As we move through the QIS process, we note that an initial component is to bring the measurement process into a state of control. In some instances, applied research will be required at this point to determine the special causes of variation affecting the measurement process. Continuing through the QIS process, applied research might also be required to identify the common causes of variability in order to reduce measurement error or bias to an acceptable or capable level.

After the measurement process is brought into an acceptable state, the process moves to assessing the control and capability of the measured product or process variables targeted for improvement. Like the work that might be required for the measurement process, applied research might be required to bring the process into a state of control, or capability, or both. In the improvement phase of the QIS,† applied research, carried out through experimental design, is a major tool commonly employed for the identification of first- and second-order critical characteristics.

3. Strategic research questions generated while using the problem-solving strategy in an effort to eliminate product or process dissatisfiers (defects, defectives, and unwanted process attributes).

Most companies employ, or have available to them, a problem-solving strategy or process in some form. Some of these strategies are the eight-step process, the four-step process, the quality improvement (QI) story, or the eight-step improvement process. All of these strategies (not to be confused with tools) employ a common set of procedures, tools, techniques, and methods. Figure 1.3 presents a generic description of the problem-solving process, based on the plan–do–check–act (PDCA) model.

Design quality may be defined at this point as the ability to properly identify those true critical quality characteristics and their appropriate target or nominal values that, if met, allow a product or process to perform in an optimal fashion based upon the stated performance and reliability requirements.

†The QIS process is fully detailed in *A quality improvement strategy for critical product and process characteristics* by J. Luftig (1989).

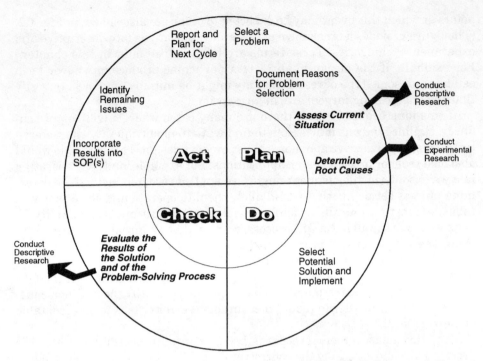

Figure 1.3 Interrelationship Between Problem-Solving Process and Research Design Applications.

As shown by Fig. 1.3, research questions are likely to arise at a number of points as a problem-solving team (PST) or natural work group employs this strategy to eliminate dissatisfiers. For example, an ingot cracking PST might assess their casting process and ultimately determine that the cast ingot cracking rate is running in a state of statistical control (this activity will be shown to be a form of descriptive research). The reduction or elimination of the cracking condition will likely depend on the determination of those conditions and causes common to the process (remember, the cracking rate is in a state of control) that are contributing to the increased likelihood of cracking an ingot during casting. One can imagine "in the old days" how such a problem might have been attacked. Using the appeal to authority, intuition, and common sense approaches, numerous expert suggestions as to how the process should be run or modified would be aggregated. Trial and error or traditional methods might be engaged to apply or test these recommendations. It is quite likely that these methods would have failed to eliminate the dissatisfier on a long-term basis. Furthermore, even if some happenstance (i.e., unplanned) associations were noted, it would be quite difficult (using the "old" methods) for the company to maintain the acquired gain by permanently modifying the body of knowledge associated with the casting process. Also "in the old days," such knowledge would probably have resided in the experiential base of one or two individuals who would eventually have come to be regarded as internal authorities. Unfortunately, when they retired, so would their insight, and problems regarded as having been eliminated would be likely to return.

The modern scientific method as a basis for industrial research significantly reduces the probability of such occurrences. Generally, the research design process model would require not only the publication of the results of the study, but also the initiation of the development or modification of standard operating procedures or practices (SOPs). The acquired knowledge and its impact on theory inherent to the organization would, therefore, be permanently captured and not rely on the presence of key personnel. With the applied research process based on the modern scientific method, the application of designed experimentation would be employed to determine the true root causes of a problem. Subsequent uses of descriptive research methods would be employed to verify the results of process changes, and to identify additional work and changes required (refer to the previously defined criteria for a research process based on the scientific method).

Whether the research opportunities arise from a technical benchmarking effort, a quality improvement effort, or a problem-solving activity, the researcher (practitioner) must now define, delineate, and delimit the research problem to arrive at a statement that may be made operational for study. This is the sequence that leads to the development of the statement of the (research) problem.

Developing a statement of the problem

The "statement of the problem" associated with a planned research study is not the same as a "problem statement," as developed for the problem-solving strategy. In planned research efforts, the statement of the problem is typically the first step in the research design process (see Fig. 1.1), and becomes the first component of the research report and executive summary. This assumes that the research study is written for internal use in industry. In academia, sections of the report preceding the statement of the problem would commonly include an introduction, a statement of the significance of the study, and possibly a section providing definitions of terms used in the study. In industry, a background statement outlining the classification of the database would have normally preceded the research design process, having been required to focus the team or individual on the research problem in the first place.

The general procedure for the development of the statement of the problem includes

1. Classification of the background data/information
2. Delineate the research problem
3. Delimit the research problem

Classification of the background data/information. This step refers to the preliminary collection, description, and statement of significance for those data that would have generated the strategic options, opportunities, or problems providing the impetus for the research study. For example, suppose that:

- As part of a customer quality assurance effort, the marketing and sales department has technically benchmarked product quality characteristics.
- The study revealed that variability in critical dimensions is significantly higher for our products than for the competitor's products.
- Insofar as our five key customers are concerned, the difference in the two populations constitutes justification for continuing to utilize us as a minor, or "filler," vendor.

The database associated with the benchmarking analysis in general, and the dimension population gap in particular, would constitute the classification base for the development of an associated statement of the research problem.

As a second example, consider the previously detailed conditions related to the ingot cracking problem. Potentially, an internal strategic product analysis as related to cost reduction may have led to ingot cracking as the most significant opportunity for improvement. These data, combined with the analysis of the cracking rate for control, would constitute the classification of the database required for the initiation of the research study and the development of the statement of the problem.

Delineate the research problem. Delineation refers primarily to the process of developing the parameters of the problem and the study. Generally, this process results in the identification of the population.*

Delimit the research problem. The object of delimitation is to arrive at a description of the research problem in a form such that a study may be effectively and correctly constructed. In an operational sense, the goal of delimitation is to develop a statement of the research problem that is concise, precise, testable, and obtainable.

These criteria may be defined as follows

Concise: A statement of the problem should be short. Its maximum length should be one or two paragraphs.

Precise: The description of the problem should be defined in detail sufficient to allow another individual, not familiar with the study itself, to correctly explain to the researcher what the study is intended to deliver.

---

*The term *population* is used at this point in reference to the target population—that is, those units, subjects, or conditions to which the researcher would like to make inferences as a result of the research effort.

Testable: The statement of the problem should be developed such that it will be possible to gather those data necessary for empirical analysis.

Obtainable: The statement of the problem should describe a research problem that has a possible answer or solution, given time, cost, and other resource constraints.

Under these conditions, the statement of the problem will allow us to properly identify the nature of the research to be conducted, and develop the resultant design procedures.

Some examples of a research problem statement appear as

1. The purpose of this study is to determine whether average axial load values (a measure of strength) for cans produced with chrome tooling are different from those for cans produced with standard steel tooling. The cans of interest for this study are NewCola and NewCola Light cans produced on all bodymakers (machines used to draw material into the shape of beverage cans) with material supplied by two suppliers: USAL and CanCo.

2. The purpose of this study is to determine the best method of packaging baked goods in order to minimize humidity incursion. Five methods of packaging will be compared for this study: wax paper boxes, metal foil boxes, plastic boxes, metal-plastic combination boxes, and the currently used wax-foil package system as a control group. The product analyzed in this study will be the CookieSnack and YummyMunch line. Humidity analysis will be analyzed with the ASTM-B123 test procedure. Packaging analysis and selection will be restricted to the currently specified Southeastern Region.

3. The purpose of this research study is to identify the potential first- and second-order critical process variables associated with plating tank no. 4 in the North plating room. The product associated with this study will be restricted to no. 12 and no. 14 cutters. Cutter quality will be assessed on the basis of chrome thickness and the absence of voids and burns.

4. The purpose of this study is to determine whether a relationship exists between the probability that a current customer has recently switched from Product A to Product B, and whether the customer has experienced product packaging failures in the past 12 months. Product packaging failures will include all characteristics currently evaluated by the product analysis group. The population included in this study will be restricted to those individuals who indicate that they purchase this product packaged as specified 80% or more of the time, and who live in a 100-mile radius of Atlanta.

5. The purpose of this study is to determine the differences in yield for four varieties of wheat across all current geographical regions and farms from which the Denver bakery currently purchases product.

6. The purpose of this study is to determine whether a relationship exists between the probability that an individual not currently recycling aluminum beer and beverage cans will decide to start recycling this product, and whether

the individual viewed the "Recycle Now" commercial during the 1996 World Series. Analysis will be restricted to those individuals in the New York metropolitan area who indicate that they did not recycle any purchased beer or beverage cans between June 1, 1995 and October 19, 1996, and who acknowledge seeing the specified commercial during the games viewed between October 19 and October 26, 1996.

As you may have already noted, the statement of the problem in a research design process is not intended to serve as an operational statement or definition or as a listing of the specific research questions to be answered. These items are detailed in subsequent stages of the design process (see Fig. 1.1). Rather, the statement of the problem at this point is intended to serve as a "stage setting" description of the nature and general purpose of the research study to be designed. Finally, as these problem statements are reviewed, it may be of interest to note that all of them may not lead to the design of an experiment. Can you categorize those that would most likely not generate an experiment? Once the classification of the database has been accomplished and presented in a background statement, and consensus has been reached on the statement of the problem, we can proceed to the selection of the type of research study to be designed.

Developing the Theoretical Framework of the Research Study

Many individuals in industry equate the term *research study* with the term *experiment*. Much of the research conducted in industry is experimental in nature. However, experimental research is only one of five major categories of research that may be conducted. Some of these categories contain types of studies or designs that are quite common in an industrial setting. Others apply more to the behavioral and social sciences and are therefore not as commonly found in industry. This volume stresses industrial research applications in general, and experimental research in particular. In order to understand the context of these studies, it is important to appreciate the total framework of research methodologies. Therefore, an overview of the many methods by which research is employed to derive knowledge is helpful.

Table 1.1 provides a breakdown of the five major forms of research and their categories of study. The first distinction presented in this table is that all research may be initially categorized as nonexperimental or experimental. Nonexperimental research design generally differs from experimental research design based on the degree to which the researcher is capable of manipulating subjects and conditions. While a researcher conducting a nonexperimental design may be capable of identifying conditions and variables of interest, he or she may not be capable of freely assigning subjects or test units to those conditions. In an experimental design, the researcher is relatively free to assign subjects or test (experimental) units to the conditions of interest (treatments) for testing. Therefore, experimental and nonexperimental

TABLE 1.1 Types of Research Designs and Their Associated Categories.

General Category	Type of Research Design	Category of Study/Inquiry
Nonexperimental	Analytical Research	Historical
		Philosophical
	Agreement Research	Consensus Analysis
		Instrument or System
		Concordance
	Descriptive Research	Status Study
		Longitudinal Study
		Case Study
		Cross-Sectional Study
	Relational Research	Concurrent Correlation
		Predictive Correlation
		Causal Comparative
Experimental	Single Factor (Comparative and Blocked)	Research Studies
	Multiple Factor	Research Studies

research differ not in the research questions posed or the conditions studied, but in the degree to which the researcher is free to manipulate conditions, treatments, and assign experimental units or subjects. Let us explore two examples of research studies that are not considered experimental in nature and that illustrate the differences between experimental and nonexperimental research.

A current topic of significant interest throughout the country relates to the issue of cigarette smoking, and particularly, its potentially harmful effects. It is not unusual to find debates raging on morning television programs related to the effect of smoking on the smoker and on bystanders. Representatives of the "smoking is harmful" position may have been heard to exclaim "Smoking causes cancer," while proponents of the opposite position have been known to defend their position by indicating "There has never been a single experiment that has shown that smoking causes cancer in humans." If we understand the basic principles of experimental design, we recognize that to some degree both sides may, in fact, be telling the truth. While the incidence of specific forms of cancer may have been shown to be statistically linked or related to whether the individual did or did not smoke, this does not mean that some other variable or factor (geographical region where the individual lives, socioeconomic status, race, etc.) might not be the true causal effect. Further, even if some external factor was not the "true" cause, a third and unknown internal factor might explain the higher incidence rate. For example, some genetic influence may give some individuals a higher potential for developing a cancerous condition, while also providing those same individuals with a tendency toward smoking, or a tendency toward nicotine addiction. If one is not careful to remember that all significant relationships may not be causal in nature, one may end up concluding that ice cream causes drowning, since drowning deaths increase in the same months of the year that ice cream sales increase.

Why not run a true experiment on smoking? Because when people (subjects) rather than coils of aluminum, cans, crop fields, packaging materials, or wheat varieties (experimental units) are involved, the researcher is not free to manipulate any and all conditions he or she might desire, and is even less capable of assigning subjects randomly to treatments. In this case, a true experiment would require that people, when babies, be assigned to one of two groups. One group would be randomly selected as those individuals who would be forced to smoke when they got older, whether they wanted to or not. The second group of individuals, when they aged, would not be permitted to smoke, again, whether they wished to or not. At the end of 30 or 40 years, the incident rates of cancer could be statistically compared. If the smoking group had a higher cancer rate than the nonsmoking group to a degree that could not be explained by chance and chance alone (statisticians and researchers refer to this as a statistically significant difference, and use inferential statistics for this purpose), then a causal influence could be inferred. It would not be reasonable to expect this experiment to be conducted in the near future. That leads us naturally to the next question: How can other groups state "Research does show that smoking causes cancer"? The answer is found by understanding that not all research is conducted by running an experiment.

Causal comparative research, also referred to as quasi-experimental research, depends upon a series of studies that, when taken as a group, provide a preponderance of evidence that can be used to inductively generate a major premise or principle. In other words, let us assume that a researcher conducts a study in the Northwest region of the country, among middle socioeconomic groups of individuals, across all ethnic groups living in that region. Suppose further that the sample group was randomly selected from this well-defined population, and categorized by whether they smoked or did not smoke. Then, suppose the same individuals were categorized as having experienced or not experienced some form of lung or respiratory cancerous condition. Finally, assume that the study showed a higher incidence rate for smokers versus nonsmokers; that is, a statistically significant relationship (notice that we have not used the word "cause" in the review of this research study).

Let us imagine, however, that the same type of study (again, note that we are talking about a study, not an experiment) has been conducted in another part of the country, by a different researcher, with different ethnic groups. Suppose further that the same significant relationship was found as in the first case. Now, imagine this scenario repeated through time, always yielding the same result. At some point, the preponderance of evidence would suggest that the observed relationship could not be intuitively explained by tertiary or nuisance variables, and therefore that a causal (and, in this case, detrimental) relationship must exist between smoking and the probability of developing some types of cancerous condition.

Let us briefly review a second example that illustrates the difference between experimental and nonexperimental research. Some years ago, one of the authors was engaged to conduct a study for the Minnesota Department of Corrections. The purpose of the study was to determine whether the provision of vocational education to inmates reduced their recidivism rate after their release. (Recidivism in this case is defined as the return to prison.) At that point in time, all inmates did not participate in this training because it was strictly voluntary. None of the other conditions or external parameters that are known to affect the probability of returning to prison after release could be controlled or manipulated by the researcher (e.g., marital status, age at first incarceration, offense, etc.). Statistically speaking, the researcher knew that these other factors could be accounted for, and "separated" from the education effect. What the researcher faced, however, was the problem of assignment. In order to run a true experiment, incoming inmates would have to be assigned to receive or not receive training during their incarceration, whether they wanted it or not. If this was not done, then the very factors that would cause an individual to volunteer would be hopelessly confounded (combined or meshed) with the reception of training. Therefore, a difference in recidivism rates attributed (on a causal basis) to vocational education could truly be attributable to those conditions or reasons that prompted the individual to volunteer in the first place. In this case, statements of causality would simply not be defensible. Only the fact that a relationship existed between a recidivistic tendency and the reception of training could be justified.

The Minnesota Department of Corrections personnel at that time, however, were not interested in relational research of a predictive correlation nature (refer to Table 1.1).

Rather than contribute another study to the preponderance of evidence base of research in this field, they offered the author the opportunity to conduct a "true" experiment. That is, they indicated that the random assignment of subjects to the treatment was an acceptable option in their minds. As with many cases in medical and behavioral research, this provides a serious ethical and moral dilemma. The review of the available literature in this field suggested that:

1. Vocational education during incarceration had been shown to be effective (in a relational rather than causal context) as a contributing factor in reducing the probability that any given individual would return to prison, but
2. The more frequently an individual is incarcerated, the more likely he or she will be reincarcerated regardless of the treatment received during imprisonment.

Conducting a true experiment in this case by randomly assigning individuals to the treatment variable (the reception of vocational education) would, if the research was correct, constitute controlling the lives of some of the test subjects. Someone who, given the opportunity to receive vocational education, would not have returned to prison, might not have had that opportunity to participate given the luck of the random draw. The subject might then be inextricably caught up in the cycle of recidivism, never succeeding in breaking the web of conditions that has proven so costly to society. What has to be weighed in this type of case is the issue of greater good. Suppose, for example, the researcher was facing the possibility that all vocational training was in danger of being eliminated for the next 10 or 20 years as a cost-cutting measure, and only an experiment that would provide evidence of a causal nature could save the programs. In such a case, "sacrificing" a few subjects for the future benefit of hundreds of individuals might justify the approach. Frequently, researchers in nonindustrial settings must face such choices. Fortunately, this was not the case in the Minnesota example. A quasi-experimental study was conducted, that, incidentally, showed a significant relationship between the two variables of interest.

In agricultural and industrial settings, we find that the same research conditions and issues exist, but generally they can be manipulated. Consider a farmer who wishes to compare two fertilizers in the context of their ability to increase barley yield. After using one of the fertilizers in one field, and applying the second fertilizer to a second field (perhaps some distance away), differences in crop yield for the two fields might be noted. However, would the difference in yield be attributable only to the fertilizer tested? Most people would recognize that the answer to this question would be, "Not necessarily." For example, the variability in yield might be due to inherent differences in the soil conditions between the two fields. The difference between this case and the smoking example, of course, is that the farmer has a choice in how to run the study. The two types of fertilizer could have been applied in some fashion to both fields. Individual and randomly selected sections or plots within each field could have been used to eliminate, control, or explain within-field variability. In Chaps. 4 and 5, these decisions, methods, and techniques are shown to be the basis for the construction of sound experimental designs. These same principles apply to industrial experiments conducted in production settings that are focused on the comparison of suppliers, process set point effects, line conditions, and other ambient control parameters. Cans, coils, ingots, connectors, circuit boards, bundles, and other experimental units of interest in industry may be freely assigned and ordered by the researcher, particularly relative to test subjects (i.e., people). Incidentally, terms that we will review such as "blocking" and "split-plot design" reveal the agricultural sources of many of the experimental design tools and approaches we use today.

Let us summarize some of the principles related to the conduct of valid research. It is important to classify the type of research study that will be conducted because this selection determines the design procedure, the method of stating the objectives of the study, and the type of conclusions, inferences, and recommendations that may be correctly provided as a result of the research conducted.

Experimental studies are generally considered to be more powerful than nonexperimental studies in finding causal relationships between and among variables (Hicks, 1964; Spector, 1982; and Wallis and Roberts, 1962). This is primarily due to the ability of the researcher to directly manipulate test units and employ principles of randomization, replication, and comparison (Natrella, 1963).

Statements and generalizations of causality may be advanced only on the basis of experimental research. Statements of causality are not permitted when resulting from a single, individual research study of a nonexperimental nature, regardless of the internal validity of the research (the degree to which the study is technically correct). Two points amplify this observation:

1. This statement is not intended as a value judgment on experimental versus nonexperimental research. As pointed out by Spector (1982), virtually all current knowledge associated with the field of astronomy is based on non-experimental research. Rather, the highest appropriate level of research design should be employed for each research problem identified, and the requirements and limitations for each type of research study must be recognized by the practitioners who perform the study, and the managers who receive its results.
2. Nonexperimental research, cumulatively, may lead to inferences of causality. The previous statement simply notes that only statements of relationship may be generated as a result of the conduct of any single nonexperimental study.

The reader should not assume that the control and manipulation of treatments, test units, and extraneous variables are uniquely associated with experimental research. That is, do not assume that when we conduct experiments, everything is manipulated, but when we conduct nonexperimental research, nothing is controlled and manipulated. In reality, as we explore and discuss the various forms of research, we find that there exists a required continuum of manipulation and control. In general, the further one moves from analytical research (see Table 1.1) toward experimental research, there is more rigor and capability to control and manipulate variables.

Having reviewed the primary distinctions between the two major categories of research, we move to an overview of five major types of research. Interestingly, four of these: analytical, agreement, descriptive, and relational are nonexperimental as illustrated in Table 1.1. Only the general category experimental has the attributes that allow the researcher to truly describe, define, or discover causal relationships.

Analytical research

According to Sax (1979), analytical research may be generally defined as research that is based on the derivation of relationships within a logical, deductive system as opposed to an empirically driven method. This type of research design includes two well-defined categories of inquiry named *historical* and *philosophical*.

Historical research is that category of analytical research involved with the retrieval of data corresponding to past events or conditions. A key aspect of this type of study is that the data are collected rather than generated; that is, no attempt is made to control the outcome. One application of historical research in industry is the "action pass" or "data sweep" analysis. A new term for this type of analysis is "data mining." In such studies, a historical database associated with a large number of first- and second-order process variables is analyzed by using complex mathematical and statistical models. The purpose of such a study is to attempt to identify potential relationships among product and process variables for further study. This type of study is often employed where (1) virtually no analysis has been conducted in the past, (2) personnel within the organization have little or no information related to "where to start" trying to solve a particular problem or understand specific sources of variation, and (3) literally no prior attempts have been made to statistically control any product or process characteristics. This type of study is distinguished from a concurrent correlation study (that will be discussed shortly) in that there is literally no manipulation or control of any variable involved in the study. This distinction, of course, is the basis for highlighting the difference between *collecting* and *studying* data and generating and analyzing data. The first approach is highly passive, the second ranges from moderately to highly proactive.

Philosophical research may be described as the attempt to establish theories or relationships when (1) no data are available as related to the statement of the problem and (2) no empirical methods are available for developing or extracting substantiated evidence (Sax, 1979). This category of study does not find extensive use in industry.

Agreement research

Agreement research may be defined as that type of research designed to obtain or create some form of consensus or consistency. Usually, this agreement is derived or generated among experts, or a group of individuals designated for study or analysis. These studies may be involved in arriving at consensus for a measurement scale, a prioritization list, or any other condition or system where a group decision is desired. The two categories of study within this area of research are consensus analysis and concordance analysis. Both of these forms of study are employed in industry, with concordance analysis employed slightly more frequently.

Consensus analysis in industrial applications includes those efforts associated with two distinct research problems:

1. To establish the degree of consistency among individuals or groups of individuals as related to a selected response or measurement scale.

2. To employ the input of a group of "experts" or individuals assigned to perform a particular task in an attempt to arrive at a final selection or prioritization list.

The first subcategory noted is related to many efforts frequently found in the conduct of market research. Often, a researcher may be economically restricted in the amount of surveying that can be conducted for a particular study. While the researcher may wish to obtain data for a conjoint analysis from all of the population purchasing a particular product, if it could be established that consistency or agreement existed between two or more of the subpopulations of interest, the study might be reduced through carefully arranged stratified sampling. Assessing and establishing the degree of consistency among the various subpopulations would be an example of the type of research problem requiring this type of research approach.

The second category of consensus analysis listed includes studies designed to create some form of prioritization or section list or scale for subsequent action. For example, those studies often employed to identify an organization's key or critical customers for partnering or surveying often fall into this category.

Concordance analyses or studies include a wide spectrum of studies in industry, often associated with measurement processes or systems. The major thrust of most studies in this category is to (1) establish the degree of concordance or (2) develop the level of concordance (as a form of consistency) between or within individuals, groups, or systems. Examples of research studies within this category include applications such as

- Establishing the concordance between a laboratory and field test
- Establishing and/or increasing the degree of the concordance of measures within an individual panel of judges or raters
- Establishing and/or increasing the degree of concordance between multiple panels of raters or judges
- Establishing the accuracy for a panel of judges, or for individual judges, inspectors, or raters, against the measures supplied by a selected "expert"

Descriptive research

Descriptive research consists of those lines of inquiry dedicated to describing current situations. Descriptive research is similar to historical research in that the researcher makes no effort to affect or influence the resultant data in any way. It differs from historical research in the currency of the data retrieved or gathered. Unlike relational research (the next category reviewed), the researcher makes no attempt to establish or discuss relationships. Simply speaking, this type of research is designed to describe a situation. There are four major categories of descriptive research, some of which find extensive use in industry:

1. A *status study* attempts to acquire data related to a situation, judgment, opinion, or attitude at a given point in time. Typically, the status study

employs a great deal of information acquired from large sample sizes and is sometimes characterized as a survey or nominative study. Customer satisfaction surveys are one form of the status study commonly found in industry. Chapter 7 provides an overview of the elements associated with obtaining valid and reliable data for this type of study.

2. A *longitudinal study* may be thought of as a status study that describes a situation or condition through time, as opposed to one point in time. For example, a research effort designed to follow graduates of a particular program through time to determine the percentage of individuals working in their field of study would be considered a longitudinal study. This type of study is not frequently found in industry. Research that follows products through time in order to assess their reliability often is conducted in industry, but in such a way so as to draw conclusions about relationships. This type of study would therefore fall under the relational, rather than descriptive, category of research.

3. The *case study* has been defined as the study of changes over time for a single individual. In industry, this term is often used to describe the study of a sequence of conditions or events as associated with a particular effort or task. For example, the accumulation of those data detailing the methods, techniques, and occurrences associated with bringing a particular production process into a state of control and capability could be generated for instructional purposes as a case study. Strictly speaking, in order to be considered an example of a case study, the determination that certain data were to be accumulated for some future illustrative purpose would be made before the activity was initiated, rather than as an afterthought.

4. A *cross-sectional* study may be thought of as longitudinal research, but where different individuals or groups are assessed for changes or differences through time. It is extensively used in educational, behavioral, and social science applications.

Relational research

Relational research includes those studies which assess the relationship between two or more variables or values. The research problem is generally focused on determining whether a relationship exists, followed by an attempt to logically explain the phenomenon. It is important to note that in this type of study, regardless of the strength of association or correlation observed, causal inferences may not be made. The statistical strength of an observed relationship has never been, is not, and never will be an indication of the probability that one variable causes (versus accompanies) a change in a second variable. There are three general categories of relational research.

Concurrent correlation research is conducted in those instances when the researcher wants to establish whether a relationship exists between two or more variables, for a single set of subjects or experimental (test) units.

TABLE 1.2 Contingency Table for Customer Change Date.

		Currently Considering Switching from Rando to Acme		
		YES	NO	
Experienced Breakage Problem During Last 12 Months	YES	72	60	132
	NO	33	35	68
		105	95	200

Consider the contingency table in Table 1.2, reproduced from the results of a marketing study conducted for the purpose of assessing customer satisfaction and dissatisfaction.

As reflected by this table, one of the research questions that the researcher was attempting to answer was

> Is there a relationship between the probability that a customer is considering switching from our brand to that of our competitor, and has the customer experienced breakage problems in the past 12 months?

Another type of research question that might justify the use of a concurrent correlation study is illustrated by the following example:

> Is there a relationship between whether we use casting filters from three different suppliers—Acme, BeeSee, or FBN—and the end-of-line porosity level of our magnesium castings?

The data set that might be collected in conjunction with this research question, and its associated statistical analysis, is presented in Table 1.3.

Predictive correlation research is concerned with establishing the relationship or degree of association between two or more variables through time, as opposed to concurrently in time. In order to understand the nature of this type of study, consider the following situation:

> Metal parts stamped by a high-quality manufacturer are examined by lot, and visually analyzed for the frequency and severity of surface flaws. Based upon current standard operating practices, the lot is assessed as "OK" or "NOT OK." The lots passing the test are then shipped to the customer, a cable manufacturer, which runs the lot of parts through an assembly line. Sometimes the lot runs completely through the assembly machine without incident; sometimes it does not. The appeals to authority and intuition that have been employed as the sources of knowledge in the past have identified surface flaws as the likely cause of line problems. The supplier has concluded that lots that had been scrapped for surface flaws would have run through the customer's line without incident. Further, the supplier, in an appeal to common sense, also believes that surface flaws could not be related to the observed line failures at the customer location.

TABLE 1.3 Contingency Table and Associated Statistical Analysis of Filter Supplier and Product Quality.

		Vendor			
		ACME	BeeSee	FBN	
End of Line Porosity	Acceptable	53	39	23	115
	Unacceptable	22	6	7	35
		75	45	30	150

Inferential statistics generated for porosity data.

Chi-Square	Value	DF	Significance
Pearson	4.02484	2	.13366
Likelihood ratio	4.27836	2	.11775
Mantel-Haenszel test for linear association	1.22890	1	.26762
Minimum expected frequency	-7.000		

Statistic	Value	ASE1	T-value	Approximate Significance
Phi	.16381			.13366 [a]
Cramer's V	.16381			.13366 [a]
Contingency coefficient	.16165			.13366 [a]
Lambda:				
symmetric	.00000	.00000		
with POROSITY dependent	.00000	.00000		
with VENDOR dependent	.00000	.00000		
Goodman and Kruskal Tau:				
with POROSITY dependent	.02683	.02397		.13547 [b]
with VENDOR dependent	.01623	.01490		.08909 [b]
Uncertainty coefficient:				
symmetric	.01813	.01678	1.07786	.11775 [c]
with POROSITY dependent	.02625	.02421	1.07786	.11775 [c]
with VENDOR dependent	.01385	.01286	1.07786	.11775 [c]
Kendall's Tau-b	-.10192	.07906	-1.28279	
Kendall's Tau-c	-.09600	.07484	-1.28279	
Gamma	-.21916	.17214	-1.28279	
Somers' D:				
symmetric	-.09818	.07616	-1.28279	
with POROSITY dependent	-.07742	.06060	-1.28279	
with VENDOR dependent	-.13416	.10367	-1.28279	
Eta:				
with POROSITY dependent	.16381			
with VENDOR dependent	.09082			

[a] Pearson chi-square probability
[b] Based on chi-square approximation
[c] Likelihood ratio chi-square probability

The customer and the supplier decide to partner on a short-term study, designed to assess (among other factors) the predictive efficiency of the surface flaw test. The next 100 lots of parts are assessed as "OK" or "NOT OK" under current standards at the supplier's site. They are all shipped to the customer site, where they are run through the assembly line. The customer records, for each lot of parts, whether the line "DID" or "DID NOT" go down during assembly for problems related to the metal parts.

This situation would lend itself well to a predictive correlation study. The researcher is interested in whether a relationship exists between two variables, where the purpose of the relationship is to allow for the prediction of one variable, by measuring the value of a second variable. Incidentally, the statistical coefficient or measure that would be employed to answer this research question would be the J index of predictive efficiency. Chapter 9 contains guidelines for the selection of appropriate inferential statistics where relationships are involved.

These types of studies are also found in industry where it is important to identify covariates that may be used for process control. An example may be illustrated by a molding process where the final characteristic of interest is fatigue strength. Suppose further that the test is not only destructive in nature, but also is extremely expensive and time-consuming to run. The process engineer and/or NWG responsible for maintaining the quality level associated with this product might be interested in finding product characteristics that (1) might be nondestructively and more efficiently tested and (2) when assessed individually or in conjunction would allow the final fatigue strength of the product to be accurately predicted. If a relationship does exist between fatigue strength and some other product characteristic, that characteristic could be tested (presumably at less cost). The fatigue strength test could then be eliminated, resulting in a cost reduction without risking a reduction in product quality.

Note that in this case, we are attempting to determine if there is a predictive relationship between one (or more) of the covariates and the final fatigue strength value. We are not attempting to explain what causes higher or lower strength values.

Causal comparative studies are also sometimes referred to as *quasi-experimental, ex post facto,* and *pseudoexperimental* studies. In causal comparative studies (recall the initial discussion of the differences between experimental and nonexperimental research), the researcher has the opportunity to manipulate all or most of the variables associated with the study with the exception of:

1. The ability to randomly assign subjects or test units to the treatments of interest or
2. The ability to fully manipulate the treatment schedule or
3. Both 1 and 2

While these studies may appear to be true experiments to the novice practitioner, statements of causality should not usually be based on this type of research study. Incidentally, for those who have read some material on this topic in the past, do not confuse a causal comparative design with the problem of finding that groups to be compared initially vary prior to treatment. This condition may be statistically managed within the context of a true experimental design.

Experimental research

As previously discussed, experimental research is the only type of research that allows the researcher to describe, define, or discover causal relationships. As Spector (1982) points out, it is important to remember that statements of causality resulting from experiments, incorrectly conducted, will be just as (if not more) inappropriate as statements of causality from relational research. The science of properly planning, designing, and conducting experiments is the primary topic for the remainder of this book. This body of content is extensive. Rather than present a description of these designs at this point, a comprehensive breakdown of the major forms of experimental designs is presented in Chap. 4, with case studies illustrating the different forms in Chap. 11.

After the statement of the problem in a research study, the specific objectives of the research are commonly detailed. This is usually done through the development and presentation of either (1) research questions to be answered or (2) research hypotheses to be tested. While the authors recognize and admit that the recommended convention that follows is not universally accepted, *research questions* should be employed for research studies of a nonexperimental nature. *Research hypotheses* should be utilized for experimental research. Guidelines for the appropriate construction of these items will be reviewed in the next portion of this chapter.

Making the Statement of the Problem Operational: Research Questions and Hypotheses

So far, we have discussed the development of the statement of the problem and the identification of the type of research to be conducted. We have already noticed that if the type of study to be conducted is nonexperimental in nature, research questions should be employed and if an experimental design is called for, research hypotheses should be used. While some may suggest that the difference between the two formats is one of convention only, it is important that the format of the operational premises of the statement of the problem is used to effectively transmit the nature of the study and the expectations regarding the analysis of the data.

When we use the term *research question*, we refer to a statement of inquiry about information or knowledge concerning a specific point. A research question does not specify an expected outcome, relationship, or difference. For example, consider the statement of the problem that was presented earlier:

The purpose of this study is to determine whether average axial load values (a measure of the strength of a beverage can) for cans produced with chrome tooling are different from those for cans produced with standard steel tooling. The cans of interest for this study are NewCola and NewCola Light cans produced on all bodymakers (the machines used to draw metal into the final shape of the can) with material supplied by two suppliers: USAL and CanCo.

If this study were to be conducted with a nonexperimental design (not an underlying requirement or assumption), then the following research questions might apply:

- Is there a significant relationship between the average lot value for axial load and tooling type (chrome and steel) on all bodymakers for NewCola cans produced with CanCo aluminum?
- Is there a significant relationship between the average lot value for axial load and tooling type (chrome and steel) on all bodymakers for NewCola cans produced with USAL aluminum?
- Is there a significant relationship between the average lot value for axial load and tooling type (chrome and steel) on all bodymakers for NewCola Light cans produced with CanCo aluminum?
- Is there a significant relationship between the average lot value for axial load and tooling type (chrome and steel) on all bodymakers for NewCola Light cans produced with USAL aluminum?

This same statement of the problem might also require the use of an experimental design. In this case, research hypotheses would be employed. According to Sax (1979), a research hypothesis is a statement of expectation regarding what the researcher believes or expects will be discovered as a result of conducting the research study. It is a tentative statement of the expected relationships between two or more variables. Research hypotheses must include both:

1. What is to be compared
2. The basis on which the comparison is to be made

Incidentally, do not confuse research hypotheses with statistical hypotheses. Statistical hypotheses are developed after the generation of the "final" experimental design, prior to the generation of the sampling plan. They are typically presented (in the report) immediately preceding the statistical analysis in the "Study" section of the research design effort. Its place in the research project sequence can be found in Fig. 1.1.

For purposes of providing a comparative example, we might find that the following statement of the problem:

The purpose of this study is to determine whether average axial load values for cans produced with chrome tooling are different from those for cans produced with standard steel tooling. The cans of interest for this study are NewCola and NewCola Light cans produced on all bodymakers with material supplied by USAL and CanCo.

could yield the following research hypotheses:

- There is no significant difference in average axial load between NewCola and NewCola Light cans produced with CanCo and USAL aluminum on all bodymakers in the plant, using steel and chrome tooling.
- NewCola cans manufactured with chrome tooling yield a significantly higher average axial load than NewCola cans manufactured with standard steel tooling.
- There is no significant difference in the average axial load for NewCola cans produced using CanCo aluminum and NewCola cans produced using USAL aluminum.
- There is no significant difference in the average axial load between all bodymakers in the plant for NewCola cans.
- NewCola Light cans manufactured with chrome tooling yield a significantly higher average axial load than NewCola Light cans manufactured with standard steel tooling.
- There is no significant difference in the average axial load for NewCola Light cans produced using CanCo aluminum and NewCola Light cans produced using USAL aluminum.
- There is no significant difference in the average axial load between all bodymakers in the plant for NewCola Light cans.

As we will discuss in Chap. 5, these research hypotheses will require that we test at least 15 statistical hypotheses.

As the developed research questions and hypotheses are reviewed, it should be noted that there are a number of attributes or criteria that may be employed to assess their sufficiency. While these elements are generally presented in the literature as related to hypotheses, they possess equal applicability to research questions.

Criteria for Evaluating Research Questions and Hypotheses

The following criteria are distilled from discussions on the subject in *Handbook of Research Design and Social Measurement* (Miller, 1964) and the seminal work on foundations of research by Sax (1979):

1. *Clarity.* Hypotheses and questions must be conceptually clear, and operationally defined where possible. Specifically, they should be stated in such a way as to effectively transmit to the audience of the study and its subsequent report exactly what is to be determined by conducting the research.

2. *Deduction of consequences.* Hypotheses and questions must be testable. General statements that do not allow for the effective application of empirical methods or statistical analyses (where appropriate) are, by definition, useless.

3. *Consistent with known facts.* Hypotheses and questions developed should be consistent with known facts and principles. In industry, this criterion should not be confused with facts obtained through the historical appeals to intuition and authority. In fact, the empirical analysis of premises obtained through these methods is well justified, and frequently beneficial. On the other hand, research hypotheses and questions should not be developed on the basis of information possessed by the solitary researcher. Previously obtained data and knowledge generated through the use of the modern scientific method should be extracted, distilled, and appropriately utilized in the development of subsequent research.

4. *Specificity.* The hypotheses and questions should be written as specifically as possible. This will generally result in multiple hypotheses or questions (as previously illustrated) for a single statement of the problem. Spector (1982) brings to our attention that many researchers use the existence of multiple hypotheses as a measure of the adequacy for this criterion. The general assumption underlying this supposition is that few problems are so simple as to require the analysis of only a single specific question or hypothesis. The presence of multiple hypotheses or questions, therefore, is taken as one indication that the requirement of specificity has been met. A second measure of specificity is the determination of whether each individual research question or hypothesis will allow for the direct development of a single statistical hypothesis. Note that research questions, as well as hypotheses, will result in the development of statistical hypotheses in conjunction with agreement, descriptive, and relational research studies.

5. *Law of Parsimony.* The law of parsimony essentially suggests that a good hypothesis or research question is stated as simply as possible, while still supported by known facts.

Self-Review Activity 1.1. Imagine that we have created a background database for the deployment of activities oriented toward the urgent need for cost reduction in a particular manufacturing facility. Financial impact data and cost-benefit analyses have revealed that one of the major costs in this facility correspond to the purchase of knives for use on a machine that slits material. These knives are used until they are worn and are replaced on a regular basis. Not only are the knives themselves expensive, but the downtime associated with their replacement is significant (mean replacement time has been calculated to be 32.5 minutes). As a result of an appeal to common sense based upon previous experience, some of the plant's line personnel have indicated that the mean time between failure (MTBF) varies for knives purchased from the three current suppliers. These data have never been assessed on a formal basis. However, a permanent record has been maintained by the production organization for the last 3 years as related to both (1) time to failure and (2) time for replacement. A decision has been made to review this database, and seek out potential opportunities.

Your assignment as related to this problem description is to:

1. Classify the type of research study most likely to be conducted.

2. Write a statement of the problem.
3. Write an appropriate set of research questions or hypotheses based upon your statement of the problem.

Self-Review Activity 1.2 Assuming that experimental research is to be conducted, write a research hypothesis, or set of research hypotheses, for the following statement of the problem:

> The purpose of this study is to determine the best method of packaging for baked goods in order to minimize humidity incursion. Five methods of packaging will be compared for this study: wax paper boxes, metal foil boxes, plastic boxes, metal-plastic combination boxes, and the currently used wax-foil package system as a control group.
>
> The product analyzed in this study will be the CookieSnack and YummyMunch line. Humidity analysis will be analyzed with the ASTM-B123 test procedure. Packaging analysis and selection will be restricted to the currently specified Southeastern Region.

Checklist for Planning Industrial Research Studies

Based upon the material presented to this point, we will begin the development of a checklist for planning research studies to be conducted in industrial settings (see Fig. 1.4). At the end of each chapter (where applicable), additional content reviewed will be added to the checklist until a complete checklist is finalized in the last chapter. A copy of the complete checklist is also presented in App. A.

Types and Purposes of the Research Study

I.   Developing the Statement of the Problem

☐ A. A comprehensive and complete background data base for the research study exists, originating from a: (✓ one)

   ☐ Strategic product-market analysis
   ☐ Technical competitive benchmarking study
   ☐ Quality improvement effort
   ☐ Problem-solving effort
   ☐ Other (specify)_____

☐ B. The background data base includes a statement of significance for the problem

☐ C. The originating problem has been properly delineated and delimited

☐ D. A specific and focused statement of the problem has been developed that is: (all should have a ✓)

   ☐ Concise
   ☐ Precise
   ☐ Testable
   ☐ Obtainable

II. Classifying the Research Study

The research design to be conducted has been identified as: (✓ one)

☐ Agreement research
   ○ Consensus
   ○ Instrument/system concordance

☐ Relational research
   ○ Concurrent correlational
   ○ Predictive correlational
   ○ Causal comparative

☐ Descriptive research
   ○ Status study
   ○ Longitudinal study
   ○ Case study
   ○ Cross-sectional study

☐ Experimental research

III. Operationalizing the Statement of the Problem

☐ If the research design is nonexperimental, research questions have been developed; or if the research design is experimental, research hypotheses have been developed

☐ The research questions or hypotheses have been evaluated and assessed as: (all should have a ✓)
   ☐ Clear
   ☐ Testable
   ☐ Specific
   ☐ Simply stated
   ☐ Consistent with known facts

Figure 1.4 A Planning Checklist for Industrial Research.

Chapter

2

Developing the Experimental Design: Concepts and Definitions

Underlying Concepts

According to Wallis and Roberts (1962), the topic of modern experimental design finds its roots in the 1935 publication of the book *The Design of Experiments,* authored by Sir Ronald A. Fisher. Since 1935, much has been added to the technology of experimental design. As one traces the development of this technology, we find a general transition from the nonscientific approach, to the scientific-traditional approach, to the scientific-modern approach to the design of experiments. Each of these stages carried the utilization of the scientific method a bit further, and expanded the ability of the researcher to add to the knowledge base under investigation.

In this chapter, the term *experimental design* will relate to the process by which certain factors are selected for study and deliberately varied in a controlled fashion, for the purpose of observing the effect of such action (Juran, Gryna, and Bingham, 1979). A similar definition is the consideration of the experimental design as the designation of a particular plan for assigning subjects or experimental units to experimental conditions, that may also include the statistical analysis of the resultant data (Kirk, 1968). In order to fully explore the implications, methods, and techniques of experimental design, two topics must first be reviewed. They are

1. Common definitions of terms that will allow for a discussion and exploration of experimental design theory
2. The basic premises associated with statistical inference.

In providing a series of basic definitions of terms associated with modern experimental design theory, one cannot help but notice the agricultural basis of many of the terms used today to describe experimental conditions. As Natrella (1963) points out, the early developments and applications of experimental design theory were in the field of agriculture, where the terms employed had an actual physical meaning. She points out, for example, that the "experimental area" was a designated piece of ground. A "block" was a smaller piece of ground, generally small enough to be considered uniform in topography, soil, and weather conditions. A "plot" was a subdivided section of a block, or a piece of ground even smaller than a block. A "treatment" was actually an agricultural treatment, such as the application of a fertilizer. "Yield" was a term used to describe the quantity of crop (e.g., wheat, barley, corn) harvested, weighed, and measured.

These terms still have direct meaning in agricultural settings. As experimental design technology was expanded into the industrial setting, these terms, as well as many others, took on a broader meaning. The expansion and modification of terms, however, did not significantly modify the essence of the terms' original meaning. For example, yield is taken today to define a measured result or outcome (Juran, Gryna, and Bingham, 1979). In chemical applications, yield will still be highly descriptive of the outcome of an experiment. In physics, the term will have more of an interpretive meaning. In the definitions that follow, the reader should note that not all authors define these terms in exactly the same manner. The definitions provided have been selected or provided in an effort to reflect a general consensus as to the meaning of each term.

Basic Definitions of Terms

Variable A variable is a quantitative or qualitative entity that can vary or take on different values. In research applications, variables are the elements that are measured and represent the concepts studied. For convenience, variables are often classified as independent or dependent to identify their relative role in the research study (Spector, 1982).

Dependent (response) variable The dependent, or response, variable is that variable which represents the outcome or end condition that the researcher is interested in affecting or understanding. It is the variable studied in terms of its value, variability, related effects, or (naturally) response. Examples of responses, the quantity being measured, include height, cleanliness, deviation from color target, and time to failure. In fact, virtually any external customer satisfier (quality characteristic) or dissatisfier (defect or defective unit) may potentially be studied as a dependent variable.

Criterion measure A criterion measure is the method or scale by which the dependent variable is quantitatively or qualitatively assessed. For example, the dependent variable in an experiment might be identified as "connector fit." This dependent variable, however, might possess numerous characteristics to be measured, analyzed, and balanced within a single experiment. Width, length, and insertion force might constitute the criterion measures associated with the dependent variable of connector fit for a given experiment. Where only a single criterion measure exists, the dependent variable and criterion measure are typically the same quantity.

Independent variable Broadly stated, the independent variable is a variable or attribute that (1) influences or (2) is suspected to influence a given dependent variable by affecting one or more criterion measures (Ford Motor Company, 1972). Independent variables may also be quantitative (e.g., temperature in degrees, time in minutes) or qualitative in nature (e.g., machine, operator, production line, plant) (Juran, Gryna, and Bingham, 1979). Independent variables, as related to experimental design theory, can provide us with some confusion if we are not careful to describe the nature of their classification. First, independent variables may be categorized as known (i.e., recognized or identifiable) or unknown. If the independent variable is known *and* included in the study, it may be described as a treatment variable. In certain types of experiments, called *factorial designs,* the treatment variables may also be referred to as factors, or main effects. Independent variables that are known, but not included in the study as a treatment variable, variables that are known but nonmanipulable, or variables that are unknown, are often referred to as nuisance variables.* One of the purposes of understanding experimental design theory is to allow for the appropriate handling of these nuisance variables. Failure to include these variables in considering the design of the experiment can lead to either inconclusive or incorrect results.

Correlation coefficient A correlation coefficient is a value that represents the relationship between two or more variables. These indices may indicate independence between the variables measured, or may suggest a relationship, positive or negative, between the elements measured, although the presence of a relationship does not guarantee that a causal association exists (Duncan and Luftig & Warren International, 1995).

Single-factor experiment An experiment with only one treatment variable.

Factorial experiment A general classification term referring to those experiments in which one or more trials or tests are conducted at every combination of the factor levels. Factorial experiments are those studies that include the study of two or more treatment variables.

Level The levels of a factor (or treatment variable, or main effect) are the values of the factor being examined in the experiment (Juran, Gryna, and Bingham, 1979). In the case of a quantitative factor, such as temperature, this might consist of studying the effects of temperature at 150, 200, and 250°. In the case of a qualitative factor, such as operator, the levels selected might consist of Bill and Jim. In the first example, temperature might also be defined as a three-level factor while in the latter case, operator might be described as a two-level factor.

Treatment combination The prescribed levels of the factors applied to a given experimental unit. In the case of a single factor experiment, the treatment combination would relate to a single level (e.g., temperature at 250°). In the case of multiple factor studies, the treatment combination would be defined by the cell condition, or the combination of levels associated with the experimental unit in question. For example, a treatment combination might be described as related to an experimental unit produced with temperature at 250°, on production line 3, by Operator 1.

---

*Some authors, such as Kirk (1968), define independent variables as those factors under the control of the researcher and all other variables as, by definition, nuisance variables. For our purposes, however, we will consider all variables not included in the research study as treatment variables to be nuisance variables, whether they are or are not under the control of the researcher.

Interaction An interaction is a relationship between two or more factors (treatments) such that changes in the treatment combinations have an effect on the criterion measure even when changes to the single factors appear to have no effect.

Experimental (test) unit Experimental (test) units are objects to which treatments are applied. These units may consist of biological entities (subjects), natural materials, or fabricated products (Juran, Gryna, and Bingham, 1979). Coils of aluminum, aluminum cans, pallets of labels or individual labels, lots of raw material, molded parts, bags of potato chips produced, or any material forming a batch or lot may be considered potential candidates for study as experimental units.

Block A block is a portion of the experimental unit or experimental environment that is more likely to be homogeneous within itself than are different portions (batches or lots) (Juran, Gryna, and Bingham, 1979). For example, experimental unit specimens (e.g., coolant samples) from a single batch of material are likely to be more uniform than the same number of specimens from different individual batches produced through time. A blocked variable refers to a manipulable independent (nuisance) variable that is not managed as a treatment variable in a study, but is statistically included in the research to retain the external validity of the research (Ford Motor Company, 1972).

Basic tools of experimental design The basic tools of experimental design include a group of methods commonly associated with the conduct of experiments based upon modern methods, generally taken to include planned grouping, randomization, and replication.

Randomization Randomization is the process of assigning experimental units to treatment conditions, or the order of testing, in a purely chance manner (Ford Motor Company, 1972). Typically, randomization is associated with three design functions:

1. Random sampling or selection
2. Random assignment
3. Randomization of order

In any properly conducted experiment, randomization is required to:

1. Eliminate bias from the experiment (Kirk, 1968)
2. Utilize statistical tests of significance (Natrella, 1963)
3. Construct probability statements about the results (Anderson and McLean, 1974)
4. Generate valid estimates of experimental error (Natrella, 1963)
5. Determine, to a certain extent, the basis for the analysis to be used with the data collected (Anderson and McLean, 1974)

It should be noted, however, that while randomization may be a basis for the design of an experiment, it does not eliminate the effects of extraneous variables. Since randomization serves to increase experimental error, it should only be used in conjunction with the management of (1) nonmanipulable or (2) unknown variables. Known, manipulable nuisance variables should, in other words, never be "handled" by utilizing randomization alone. As we proceed to a discussion of the logic of experimental design, and the consignment of error in the study, we will review this principle extensively.

Planned grouping Planned grouping refers to those methods associated with improving the sensitivity of experimental designs by the use of specially designed experimental patterns (Natrella, 1963). Examples of these patterns include block designs, matched pairs designs, and repeated measures designs.

Replication Simply expressed, replication involves the acquisition of two or more observations under a set of identical experimental conditions, or the collection of multiple observations at each treatment combination tested (Kirk, 1968). The primary purpose of this tool is to provide a quantifiable and accurate measure of the precision of the design where precision relates to the concept of experimental error (Spector, 1982).

Experimental error Experimental error is a measure that includes all uncontrolled and/or unexplained sources of variation affecting a particular score (Kirk, 1968; Spector, 1982). Another term used to describe this quantity is *unexplained variation*.

Type I error A Type I error is an incorrect decision to reject the null hypothesis when it is really true. The probability of making a Type I error is called *alpha* (Duncan and Luftig & Warren International, 1995).

Type II error A Type II error is an incorrect decision to accept the null hypothesis when it is really false. The probability of making a Type II error is called *beta* (Duncan and Luftig & Warren International, 1995).

Inference space The inference space refers to the population or universe for which the results of the experiment may be generalized.

Population The population is the collection of all observations identified by a set of rules or boundaries (Kirk, 1968). Also referred to as a *target population* or *universe*, the term *population* refers to the aggregate of all observations of interest to the researcher (Sax, 1979). In this case, the term *of interest* refers to the total set of observations in which the researcher is interested in generalizing the results of the research study. Given this definition, it is possible to refer and generalize to a population of responses, scores, or traits (Spector, 1982).

Research population The research population is that portion of the population or universe available for sampling.

Sample A sample is a limited number of elements selected from a research population to be representative of that population. Note that the term *representative* does not imply that the sample is a miniature population in the sense of possessing identical descriptive attributes (Sax, 1979).

Representative sample Samples that have been drawn or selected from a research population in a random, unbiased fashion are referred to as *representative samples*. Random sampling refers to the process of selection whereby all samples of size n have an equal probability of occurrence (based upon probabilistic rather than purposive sampling) (Kirk, 1968; Sax, 1979; Spector, 1982). The concept of bias is also critical to this approach. Biased samples consistently overestimate or underestimate population parameters. Unbiased samples tend, in the long run, to neither overestimate or underestimate population parameters. It is absolutely essential to note that increasing the sample size as associated with unbiased samples tends to provide a more accurate representation of the research population. Increasing the number of cases associated with a biased sample, on the other hand, will not increase the accuracy of the estimate. In essence, increasing the number of observations cannot offset or compensate for the faulty selection of data in the design (Spector, 1982). As it relates to the design of the sampling plan, there is no principle associated with experimental design and the appropriate use of inferential statistics that is more critical than randomization.

Parameter A parameter is a measure computed from all observations in a population. Population or universe parameters are designated by Greek letters (Kirk, 1968).

Statistic A statistic is a measure computed from observations in a sample, designated by Latin letters (Kirk, 1968). Descriptive statistics are utilized as point estimates for population parameters.

Statistical inference Statistical inference is the process by which the known characteristics or attributes of a sample are generalized to the unknown characteristics of a population or universe (Spector, 1982). Statistical inference, properly conducted, provides (1) an estimate of the population parameter and (2) an estimate of the amount of error which may be expected (i.e., the uncertainty of the estimate) (Sax, 1979).

Control *Control* is one of those terms that is defined and described in virtually innumerable ways in the literature. It has been used to describe (1) the manipulation of subjects or conditions, (2) holding conditions or subjects constant, (3) the structuring of an investigation, or (4) the manipulation of data. For our purposes, we will use the term control to describe the process of limitation, the technique of holding a particular nuisance or extraneous variable constant at a single level.

Confounding Confounding occurs when the effect of an extraneous, nuisance, or unknown variable or effect becomes intermingled with a treatment variable (Ford Motor Company, 1972). Basically, confounding is a term used to describe the effects of a flaw or error in the experimental design. This is because the basic logic of experimental design theory is particularly concerned with preventing variables not of interest to the researcher from becoming directly related to treatment variables (Spector, 1982). This condition may give the false appearance of a treatment effect or hide the presence of a true treatment effect (Spector, 1982). Finally, note that many authors and researchers use the term *systematic error* to describe the effect of confounding. As we proceed with a subsequent illustration of the purposes of an experimental design, we will provide a number of examples of this phenomenon.

Experimental design notation Campbell and Stanley (1963) created a system of notation for experimental designs that allows for the description of various nonexperimental, experimental, and quasi-experimental (causal comparative) designs. In this system, O's are used to designate observations or measurements of variables. X's are employed to designate the application of a treatment. R is used to indicate or reflect the use of randomization, generally as related to the assignment of experimental units to groups. As an example of the use of this nomenclature, the true (Campbell and Stanley, 1963) experimental design referred to as a *pretest-posttest control group design* may be displayed as

$$R \quad O \quad X \quad O$$
$$R \quad O \quad \quad O$$

This nomenclature is particularly helpful with the display and explanation of basic, single-factor, randomized comparative designs. As the experiments increase in complexity, this nomenclature becomes somewhat less useful, and we will resort to alternative systems of nomenclature.

Internal validity Internal validity is a function of the degree to which the design is technically correct and sound, as well as the successful elimination and control of systematic error (Spector, 1982). Experimental research that is internally valid usually allows us to generalize the sample results to the equivalent effects for the research population (Campbell and Stanley, 1963).

External validity External validity refers to the overall ability to generalize, or use the results of the study (Spector, 1982). Specifically, a study is said to be externally valid if it is (1) internally valid, or technically sound, and (2) the results of the study as associated with the research population can be generalized to the target population or universe (Campbell and Stanley, 1963).

Supporting Definitions

The following definitions related to industrial applications of applied research will be useful as we explore appropriate methods for industrial research designs in later chapters.

Statistically in control A process is said to be *statistically in control* if, when measured statistically, the process consistently produces only variations that can be attributed to common causes. The presence of this situation is generally judged through the use of a statistical process control (SPC) chart (Duncan and Luftig & Warren International, 1995).

Statistically out of control A process is said to be *statistically out of control* if, when measured statistically, the process produces not only variations that can be attributed to common causes but special causes of variation as well (changes to the process that are not due to chance alone). As with statistical control, the presence of this situation is generally judged through the use of a statistical process control chart (Duncan and Luftig & Warren International, 1995).

Process capability Process capability is the degree to which the process (output) meets (or will meet) specifications or requirements on an ongoing basis. *Capability* refers to the future performance of the process, not to its past or present performance, although past or current data are used in making the assessment of capability. The prediction of future performance requires stability and predictability in the process output. Therefore, actual capability cannot be assessed until the process has been documented to be in a state of control (Duncan and Luftig & Warren International, 1995). There are three commonly accepted capability indices: C_p, C_{pk}, C_{pm}. These calculations provide an indication of the degree to which the process output is expected to meet the specifications or a target. In general, a process is considered capable if the capability indices are greater than 1.0.

Standard operating procedures (SOPs) A standard operating procedure is a written document that describes the methods and procedures a worker uses to run an operation. Manufacturing SOPs typically include descriptions of the process along with procedures that outline the following: power-up, setup, shutdown, operation, process control, adjustment, documentation, housekeeping, routine maintenance, simple troubleshooting, and diagnosis. There are also SOPs for nonmanufacturing processes (e.g., an SOP for filing a medical claim) that do not include power-up, maintenance, etc. but still outline the steps for the process (Duncan and Luftig & Warren International, 1995).

Chapter 3

An Introduction to Statistical Inference

Inferential Statistical Methods

While this text is not intended to emphasize inferential statistical methods, it is necessary to understand some of the principles associated with this topic if an understanding of experimental design theory is to be acquired. Some concepts, such as blocking, are virtually impossible to fully comprehend or appreciate without this content. We will, therefore, spend some time on an understanding of the purpose and structure of inferential statistics.

Certain assumptions and limitations should be noted before we begin. The content that follows is premised on the assumption that the reader has mastered an understanding of the random sampling distribution (RSD) theory, the normal and t distributions, and the concept of point and interval estimation. For purposes of simplicity, we will limit our discussion to the topic of mean (average) differences. The reader should note, however, that all of the concepts discussed may be extended to testing for dispersion differences, differences in rates or counts, or associative relationships. Finally, we will limit our examples (at this point) to the single factor, completely randomized design.

In order to provide a conceptual context for our discussion, we will begin our review with an example of a statement of the problem and research hypothesis.

Statement of the problem:

> The purpose of this study is to determine whether packaging our CookieSnack cookies in a new foil overlay package will resist humidity incursion as well as our current wax paper package. Humidity analysis will be assessed utilizing the currently standard ASTM-B123 Laboratory Test.

Research hypothesis:

The new foil overlay package proposed for use in packaging our CookieSnack cookies will allow significantly lower levels of humidity to penetrate the package as compared to our current wax paper package.

We will assume that an appropriate experimental design has been constructed, and that the sampling plan and testing process were correct and properly conducted. We will imagine that the humidity level data were collected, and appeared as follows:

Packaging Method (Treatment)

Foil overlay	Wax paper (control)
1.25	1.34
1.18	1.42
1.21	1.37
1.24	1.41
1.19	1.40
1.20	1.39
1.20	1.46
1.19	1.44
1.18	1.42
1.20	1.43

Using a standard statistical software package (SPSS), we can easily generate applicable descriptive statistics for the two groups of data. The result of this effort is presented in Fig. 3.1.

Typically, a graphical display or illustration of the data of interest should accompany the descriptive data set. This procedure facilitates the visualization of the raw data and provides an illustrative display of the comparisons we intend to test. Figure 3.2 presents a comparative display of the two sample data sets in the form of histograms. Figure 3.3 presents a box-and-whisker plot of the same data.

We may note from the descriptive statistics calculated that there is an

SPSS for MS WINDOWS Release 6.1

Number of valid observations (listwise) =   10.00

Variable	Mean	Std Dev	Valid Minimum	Maximum	N	Label
FOIL	1.20	.02	1.18	1.25	10	Foil Overlay
WAX	1.41	.03	1.34	1.46	10	Wax Paper (Control)

Figure 3.1 Descriptive Statistics For Humidity Levels Gathered for the Sample Packaging Experiment.

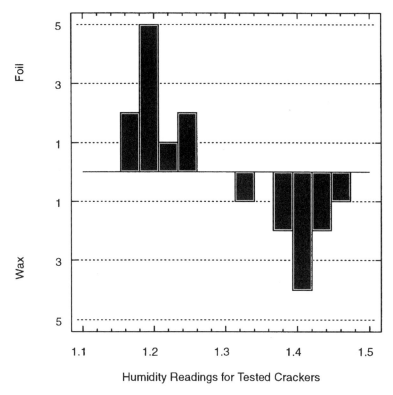

Figure 3.2 Comparative Histograms For Humidity Incursion Levels by Package.

observed difference between the average (mean) humidity levels for the experimental units (crackers) tested. Those units randomly sampled from the foil package group reflected a mean of 1.20. The units randomly drawn from the current wax paper (control) group had a sample mean of 1.41. Obviously, the two means are unequal, reflecting a calculated difference of 0.21. The issue at hand is not whether the *sample* means are different; they are. The issue is whether the observed difference is large enough to infer that the two *populations* are unequal, based upon the samples tested. This is exactly the role played by inferential statistics. That is, to determine whether an observed difference is due to chance or a true difference in the population means (for this case).

Let us examine how the utilization of inferential statistics allows us to accomplish this task. Imagine for a moment that a population exists with both:

- A mean μ of 100
- A standard deviation σ of 5

Next, suppose that we draw a random sample of size $n = 10$ from this popu-

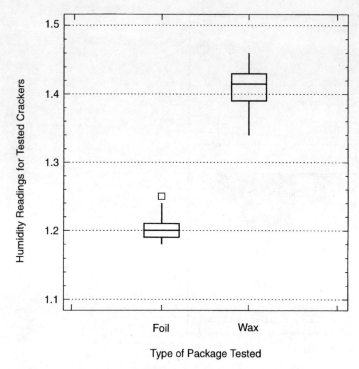

Figure 3.3 Box-and-whisker Plot for Humidity Incursion Levels by Package.

lation. We would not necessarily expect to obtain a sample mean of 100, because of sampling error. With this type of example, we define sampling error as the expected and quantifiable discrepancy between a sample statistic and its associated process or population parameter due to the variability of the population and the sample size employed. As you will recall, we are capable of quantifying the expected sampling error about the population mean with the standard error $SE_{\bar{x}}$ of the estimate, where the $SE_{\bar{x}}$ for the RSD of the means is equal to:

$$SE_{\bar{x}} = \sigma_{\bar{x}} = \frac{\sigma}{\sqrt{n}}$$

For our theoretical distribution, this value would be calculated as:

$$SE_{\bar{x}} = \sigma\bar{x} = \frac{5}{3.16} = 1.58$$

Theoretically, the RSD of the means may be approximated by the normal distribution, with (in this case) a mean of 100 and a standard deviation (our calculated standard error estimate) of 1.58. Figure 3.4 illustrates the appearance of this distribution.

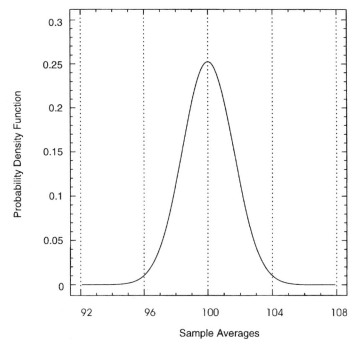

Figure 3.4 Random Sampling Distribution (RSD) for Sample Averages.

Discussion question: How might the conditions presented by this illustration be interpreted?

Given that we are attempting to compare two groups, or levels, we will take our discussion of RSD theory one step further.

Assume that:

- We have two independent populations from which we are going to draw samples of size 10 (i.e., one sample from each population).
- Each population has a mean μ of 100, and a standard deviation σ of 5.
- Both populations are normally distributed.

Now, if we consider the possible outcomes of such an exercise, we can imagine that a random sample drawn from the first population (with a mean of 100) might yield a sample average that is not equal to 100 and might not be equal to the sample average drawn at random from the second population (that also has a mean of 100).

Suppose we were to test the null statistical hypothesis that

$$H_0: \mu_1 = \mu_2$$

We would recognize, as associated with this comparison, that the expected value of the difference in the sample averages would be 0.0 if H_0 was true, but we would expect, because of sampling error, that there exists an expected and quantifiable distribution of differences in the sample averages due to the sample sizes employed and the variability inherent to the two populations.

Fortunately, statisticians have provided us with a number of formulas for calculating the standard error of the estimate for the RSD of differences between (two) sample averages. For the condition where:

1. The populations are independent
2. The populations are normally distributed
3. The populations have known and equal variances

then the standard error of the RSD would appear as:

$$\sigma_{\bar{x}} = \sqrt{\frac{\sigma_1^2}{n_1} + \frac{\sigma_2^2}{n_2}}$$

For our theoretical example, this value would be calculated to be 2.236. Using this value, we could approximate the theoretical RSD of the mean differences from these two independent populations as shown in Fig. 3.5.

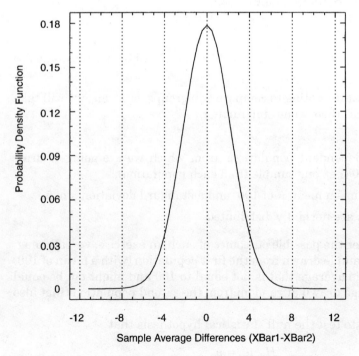

Figure 3.5 Random Sampling Distribution (RSD) for Difference Between Sample Averages.

Discussion question: Examine the RSD in Fig. 3.5. What interpretations might we make, based upon this illustration?

Now, let us move to the final stage of our discussion and return to the subject of our packaging experiment. You will recall that the data collected appeared as:

Packaging Method (Treatment)

Foil overlay	Wax paper (control)
1.25	1.34
1.18	1.42
1.21	1.37
1.24	1.41
1.19	1.40
1.20	1.39
1.20	1.46
1.19	1.44
1.18	1.42
1.20	1.43

In this case, we wish to test the hypothesis of equality for the population means, but the population standard deviations are unknown and the sample sizes are small.

In a case such as this, we use a t distribution to quantify the sampling error associated with the differences between the sample averages. The t distribution is described by the number of degrees of freedom associated with the comparison. In this case, the applicable t distribution corresponds to $(n_1 + n_2 - 2)$ df, which is $(10 + 10 - 2)$ or 18 degrees of freedom. In order to use this RSD, we must transform the observed difference in the sample averages or means into t values. These t values are based upon the number of standard error units that the observed difference falls away from the expected value of 0.0 if H_0 is true. This transformation is achieved through the application of the t-test statistic, calculated as:

$$t = \frac{\overline{X}_1 - \overline{X}_2}{\sqrt{\frac{s_p^2}{n_1} + \frac{s_p^2}{n_2}}}$$

where s_p^2 represents the pooled, or average, sample variance. This test statistic may also be conceptually represented as

$$t = \frac{\text{observed difference}}{\text{unexplained variability}}$$

or

$$t = \frac{\text{treatment effect}}{\text{experimental error}}$$

If H_0 is, in reality, true, then the expected value of t is 0.00. The larger the observed difference obtained, as compared to the experimental error present in the experiment (or model), the larger the t value. As the t value increases, the probability that the two samples were randomly drawn from populations with equal means is reduced.

This effect may be illustrated by reviewing the RSD for this t test presented in Fig. 3.6.

Using SPSS for an analysis of the two groups, we obtain the output presented in Fig. 3.7. Note the result of the hypothesis test of the equality of the means at the bottom of the figure.

As shown by this analysis, the probability (p value) that the observed difference is due to sampling error, and sampling error alone, is 0.000. We would, therefore, accept the alternative statistical hypothesis:

$$H_1 : \mu_1 \neq \mu_2$$

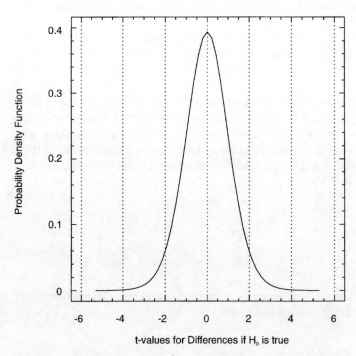

Figure 3.6 Random Sampling Distribution (RSD) for Mean Differences Based On t Distribution with 18 df.

SPSS for MS WINDOWS Release 6.1

t-tests for Independent Samples of PACKAGE   Packaging Method

Variable	Number of Cases	Mean	SD	SE of Mean
HUMIDITY				
FOIL OVERLAY	10	1.2040	.024	.007
WAX PAPER (CONTROL)	10	1.4080	.035	.011

Mean Difference = -.2040

Levene's Test for Equality of Variances: F= 1.175 P= .293

Variances	t-test for Equality of Means				95%
	t-value	df	2-Tail Sig	SE of Diff	CI for Diff
Equal	-15.30	18	.000	.013	(-.232, -.176)
Unequal	-15.30	15.83	.000	.013	(-.232, -.176)

Figure 3.7 Two-Sample Analysis Results for the Sample Packaging Experiment.

thereby rejecting the null hypothesis. We would further infer that the average humidity level for the population of experimental units corresponding to the new package (foil) material is less than the population mean for the experimental units packaged with our current material (wax paper). Note that at this juncture, we would attribute the inferred difference in the population means to the effect of the packaging material or treatment. This inference will only be correct if the experimental design is devoid of systematic error, or a confounding effect. If this is not the case, you must absolutely recognize that the statistical analysis and inference *is still valid,* while our conclusion or inference as to the *cause* of the difference will be incorrect.

This observation should lead you to suspect, correctly, that an appropriate statistical analysis of the data will not, and cannot, correct a flawed experimental design.

Before we move into a discussion of experimental design theory, a second example of a simple comparative experiment must be reviewed. In the last example, we illustrated the statistical analysis for a single factor experiment, where the treatment variable consisted of two levels. Obviously, if we carefully review the numerator of the *t* test statistic, we could not use this analytical approach if there were more than two levels. Consider the following statement of the problem for a second completely randomized design:

> The purpose of this experiment is to determine whether the flatness of beer bottle neck labels produced at the Fayetteville plant varies with the paper vendor supplying the raw stock.

This statement might yield the research hypothesis:

> There is no statistically significant difference in the performance of paper purchased from our current four vendors as measured by end-of-line (EOL) flatness for labels.

In this example, the treatment variable, or factor, is vendor. The dependent variable could be identified as EOL label quality, with flatness utilized as the criterion measure studied. If all four vendors are included in the experiment, then there would obviously be four levels associated with the study. The t test could not be used in the analysis of these data, therefore, in that we would want to test the statistical hypotheses that:

$$H_0: \mu_1 = \mu_2 = \mu_3 = \mu_4$$

$$H_1: \mu_1 \neq \mu_2 \neq \mu_3 \neq \mu_4$$

In cases such as this, we would employ the one-way analysis of variance (ANOVA) as a statistical procedure. While a comprehensive review of this procedure is beyond the scope of this text, it is important for the reader to understand and appreciate the structure and output of this analytical procedure. Further, we will be able to examine this procedure as a conceptual (as well as mathematical) extension of the t test, with similar implications as related to experimental design theory. Suppose that the coded data collected in conjunction with this experiment appeared as displayed in Table 3.1.

TABLE 3.1 Label Flatness Data Collected from the Comparative Experiment for Paper Vendor Differences.

Treatment Levels (Vendor)			
Acme	Wilson	FBNI	Akron
12	14	22	12
11	13	26	10
11	12	28	11
11	14	22	11
12	13	24	12
11	13	26	12

The One-Way Analysis of Variance (ANOVA)

The one-way ANOVA, in spite of its somewhat misleading name, is actually a test of differences in the means among more than two groups (i.e., levels). One-way ANOVA is a standard procedure employed for analyzing one factor, with $J > 2$ levels. This analytical method will allow us to test the hypotheses:

$$H_0: \mu_1 = \mu_2 = \cdots = \mu_J$$

$$H_1: \mu_1 \neq \mu_2 \neq \cdots \neq \mu_J$$

where J = total number of groups (treatment levels) tested (four in the vendor example)
j = each of the J groups
n = number of observations in each of the j groups
i = each of the individual observations in the j groups

The one-way ANOVA model allows us to compare the J group means by comparing the variation between the groups to the variation within the groups and using the F test statistic to test the significance of the effects. We may express, conceptually, this ratio as

$$F = \frac{\text{observed difference (variability) in the sample means of the treatment levels}}{\text{unexplained variability within the levels}}$$

or

$$F = \frac{\text{explained variability}}{\text{variability within the levels}}$$

or

$$F = \frac{\text{treatment effect}}{\text{experimental error}}$$

Note that this is, in essence, identical to the t test we previously used, where the difference between the two level means (the numerator of the test statistic) was divided by the variability within the two levels (the denominator of the test statistic). Further, paralleling the condition noted for the t test, the larger the observed difference between the level means (as related to the experimental error), the larger the F value becomes. The larger the F value gets, the lower the probability that the J (four in this example) samples were randomly selected from populations with equal means.

All other considerations as related to this analysis are of primarily mathematical interest. They are, therefore, for the purposes of this text, not of significant interest at this point.

TABLE 3.2 One-Way ANOVA Table.

Source of Variation	Sum of Squares (SS)	df	Mean Square	F
Treatment Effect (Between Level Effect)	$n\sum_{j=1}^{J}(\overline{X}_j - \overline{X}_J)^2$	J - 1	$\dfrac{SS_B}{df_B}$	$\dfrac{MS_B}{MS_W}$
Experimental Error (Within Level Variability)	$\sum_{j=1}^{J}\sum_{i=1}^{n}(X_{ij} - \overline{X}_j)^2$	Jn - J	$\dfrac{SS_W}{df_W}$	
Total Variation	$\sum_{j=1}^{J}\sum_{i=1}^{n}(X_{ij} - \overline{X}_J)^2$	Jn - 1		

The one-way ANOVA is generally structured in a standard table, as shown in Table 3.2.

For our experiment, the analysis of the data begins with the generation of the descriptive statistics for the four vendors (levels of the treatment vari-

SPSS for MS WINDOWS Release 6.1

For Vendor = ACME,
Number of valid observations (listwise) =   6.00

Variable	Mean	Std Dev	Valid Minimum	Maximum	N	Label
FLATNESS	11.33	.52	11.00	12.00	6	VENDOR = ACME

For Vendor = WILSON,
Number of valid observations (listwise) =   6.00

Variable	Mean	Std Dev	Valid Minimum	Maximum	N	Label
FLATNESS	13.17	.75	12.00	14.00	6	VENDOR = WILSON

For Vendor = FBNI,
Number of valid observations (listwise) =   6.00

Variable	Mean	Std Dev	Valid Minimum	Maximum	N	Label
FLATNESS	24.67	2.42	22.00	28.00	6	VENDOR = FBNI

For Vendor = AKRON,
Number of valid observations (listwise) =   6.00

Variable	Mean	Std Dev	Valid Minimum	Maximum	N	Label
FLATNESS	11.33	.82	10.00	12.00	6	VENDOR = AKRON

Figure 3.8 Descriptive Statistics for Flatness Levels Gathered for Four Paper Vendors.

```
----- ONEWAY -----
```

Variable FLATNESS
By Variable VENDOR   Paper Vendor

Analysis of Variance

Source	D.F.	Sum of Squares	Mean Squares	F Ratio	F Prob.
Between Groups	3	741.7917	247.2639	134.2609	.0000
Within Groups	20	36.8333	1.8417		
Total	23	778.6250			

Figure 3.9 One-Way Analysis of Variance (ANOVA) for Flatness Levels by Paper Vendor.

able). This analysis (accomplished through the use of SPSS) is presented in Fig. 3.8.

The next step would be to generate the one-way ANOVA table. The output produced by SPSS is presented in Fig. 3.9.

Self-Review Activity 3.1 Examine and review the results of this analysis. What conclusions might we draw? Would the null hypothesis be rejected or accepted? Describe the results of the study in terms of EOL label flatness as related to the paper vendors analyzed. What analysis would you suspect might be conducted next?

Additional information associated with inferential statistics, and the analysis of population parameters other than means, is presented in Chap. 9. At this point, we are ready to return to the topic of experimental design theory.

Chapter 4

Developing the Experimental Design: Defending the Design Against Internal and External Threats to Validity

The Basic Logic and Purpose of an Experimental Design

As we have previously discussed, the purpose of experimental design is to arrange a treatment in such a way that its effect on selected experimental units may be assessed and generalized to a predetermined population (Wallis and Roberts, 1962). While this may be the intention of the researcher, however, the planned study is not always successfully executed. This is primarily because of underestimation of the difficulties associated with properly planning and executing an experiment.

The pitfalls associated with designing experiments that are, at worst, seriously flawed and, at best, inefficient, are complex and may be structured in a number of ways. Virtually all of these threats relate to the general issue of extraneous variables. These variables are those manipulable and nonmanipulable independent variables or conditions that, if not properly considered and managed by the researcher, could affect either the dependent variables or the experimental error estimate (Sax, 1979). Proper and efficient experimental designs are those planned comparisons which allow the researcher to correctly assess the relationship between the dependent variables and the treatments, without the influence of the extraneous variables (Kirk, 1968). Consider that this goal implies that we have two objectives:

	Final Conclusions	
	The treatment(s) is related to/does affect the dependent variable(s)	The treatment(s) is not related to/does not affect the dependent variable(s)
The treatment(s) is related to/does affect the dependent variable(s)	Affect properly assessed *(Power)* / Affect properly identified but underestimated	Incorrect result due to: (a) confounded denominator; or (b) a lack of sensitivity; or (c) confounding or other threats to internal validity *(Type II error)*
The treatment(s) is not related to/does not affect the dependent variable(s)	Incorrect result due to confounding or other threats to internal validity *(Type I error)*	Correct result for the right reason(s) *(Confidence)*

Reality (Truth)

Figure 4.1 Potential Outcomes Resulting from Experimental Design Efforts.

1. If the treatment has an effect on the dependent variable, we want to detect it and properly assess its degree of influence.
2. If the treatment variable has no effect on the dependent variable, we want to detect this condition as well.

Given these objectives, there are four possible outcomes that may result from any given experiment. Figure 4.1 illustrates these potential conditions. The outcomes and their causes are presented in a somewhat simplistic but sufficient fashion to provide a framework for our discussion.

Type I error: The treatment is not related to/does not affect the dependent variable, but the experiment causes the researcher to infer that a relationship/effect does exist

We will structure our discussion of pitfalls and errors as associated with experimental designs based upon those conditions displayed in Fig. 4.1. We will begin our discussion with the cell associated with the following outcome:

> The treatment is not related to/does not affect the dependent variable, but the experiment causes the researcher to infer that a relationship/effect does exist.

This outcome, called Type I error, may be a function of (1) employing a design not truly experimental in nature or (2) incorrectly using experimental designs, allowing extraneous manipulable and nonmanipulable variables to become confounded with the treatment. This confounding has been previously described as a primary source of systematic error. Let us examine how these conditions might occur.

Pre- and quasi-experimental designs. A number of research designs for studies often used in industry are thought to be experimental designs, but, in fact, are not. In order to have a true experiment, there must be a capability to provide some form of comparison between an experimental unit receiving the treatment and an experimental unit not receiving the treatment. Therefore, a meaningful experiment must consist of two or more conditions (Kirk, 1968; Sax, 1979; Siegel and Castellan, 1988; Wallis and Roberts, 1962). Given this requirement, let us review two research designs which cannot properly be referred to as *experimental designs* and will yield incorrect conclusions if the distinction is not made. Note that the conclusions, not the results, of the research study would be incorrect if these designs are misconstrued to be experiments. This observation implies that a statistical test may yield a correct inference, but the reason for the difference might be misplaced.

The one-shot case study. The one-shot case study is, perhaps, the weakest and most insufficient research design found masquerading as an experiment. Imagine the following situation:

> Steel flat stock is being produced at a particular factory. One of the more expensive consumable materials used for this production process is the purchased coolant. Purchasing has identified a potential new vendor for this material, and has asked Operations to test the new coolant to assess its effect on the quality of the steel stock rolled. We will assume that flat stock quality (the dependent variable) is currently assessed by the ability to roll to target for thickness and profile (the two criterion measures).
>
> Deviation from target thickness and profile are not currently measured. The new coolant is brought in, run through the mill, and a number of coils are produced. For each coil, thickness deviation and profile are measured, found to be acceptable, and the coolant is accepted. The nomenclature for this study may be illustrated as

$$X \quad O$$

One hardly knows where to begin in describing all of the problems associated with the possible conclusion that the coolant will either have no deleterious effect on the quality of the material, or (worse) that the quality has been improved. Let us deal with the most obvious insufficiency first. In the absence of comparative data, an inference of improvement is simply not justified. In the absence of comparative data, there is no way to determine the level of quality before the new coolant was introduced, so there is no way to assess its change. Therefore, an inference of improvement is simply not justified.

Some personnel might believe that this study might, on the other hand, defend the assertion that future use of the coolant will not adversely affect quality. Again, you should be able to generate an extensive list consisting of all the reasons this might not be true. For example, were the measurement devices used during this period of time calibrated? Are the measurement systems in a state of statistical control? Is there a degrading effect for this coolant that does not exist with the current coolant? Were the coils utilized in the test randomly selected from a population of all coils rolled on this mill? Were the coils used representative of the target population or universe? Were the mill operating conditions standardized during this test, where the critical process variables represented the conditions under which the coolant is expected to perform? None of these questions are answerable given this type of research study. The one-shot case study might, under certain circumstances, yield data that may be valuable insofar as providing a foundation for further analysis, but it is not an experiment, and causal effects cannot be justified from such an approach.

The one-group pretest-posttest design. The one-group pretest-posttest design is thought by some individuals to abrogate or offset the deficiencies of the one-shot case study. While providing a comparative analysis, this design is still not classifiable as an experimental design. The nomenclature for this design appears as

$$O \quad X \quad O$$

In this type of research study, data or observations are gathered prior to and following the application of the treatment. The results are statistically compared, and (all too frequently) documented differences or relationships are concluded to have occurred due to the treatment. Suppose we encountered the following conditions:

> An engineer has been tracking a production process that has been running in a state of control for the last 4 to 5 months. The variable of interest relates to cable quality, where the criterion measure of interest is the length associated with the manufactured product.
>
> A new cutter design has been recommended for use with the plant's production line. The engineer draws a random sample of 15 cables from the line, and measures the length of each experimental unit. She subsequently installs the new cutters, starts up the production line, and draws a second random sample of 15 cables from the postchange production lot.

The statement of the problem generated for this study was proposed as

> The purpose of this study is to assess the effect of a new proposed cutter design on the ability to achieve a lower average deviation from target value for length on all cables manufactured at the Dallas plant.

The research hypothesis developed for this statement of the problem was stated as

> The average deviation from target on length for cables manufactured with the new cutter design will be lower than for cables manufactured with the existing (old or before) cutter design.

The statistical hypotheses tested were written as

$$H_0: \mu_{OLD} = \mu_{NEW}$$

$$H_1: \mu_{OLD} \neq \mu_{NEW}$$

The coded data collected for this study, measured in terms of deviation from target length, appeared as follows:

Old design	New design
0.35	0.42
0.34	0.44
0.32	0.48
0.33	0.45
0.35	0.44
0.32	0.47
0.31	0.44
0.36	0.41
0.30	0.50
0.39	0.40
0.35	0.46
0.34	0.48
0.37	0.42
0.39	0.41
0.35	0.45

The statistical analysis of these data is provided in Figs. 4.2, 4.3, and 4.4.

As a result of the analytical procedure, the engineer conducting the research rejected the null hypothesis, accepted the alternative hypothesis, and concluded that, in fact, the new cutter design actually increased average deviation from target thereby increasing the ability of the plant personnel to produce cables at or close to the target length.

While the statistical analysis was correct, the conclusion that the treatment had a negative effect on the criterion measure may not be. This is almost exactly the type of situation encountered by Anderson and McLean (1974), presented in the Introduction of this book. There probably was a difference in length deviation between the two production run conditions. Further, there probably was an increase in the deviation from target for length between the two tested periods. The problem is, this shift may or may not have had anything to do with the change in cutter design!

The fact of the matter is, when we use the applicable formula:

$$t = \frac{\overline{X}_1 - \overline{X}_2}{\sqrt{\dfrac{s_p^2}{n_1} + \dfrac{s_p^2}{n_2}}}$$

we believe we are actually testing:

Chapter Four

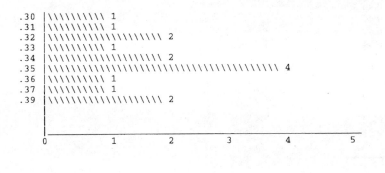

Value	Frequency	Percent	Valid Percent	Cum Percent
.30	1	6.7	6.7	6.7
.31	1	6.7	6.7	13.3
.32	2	13.3	13.3	26.7
.33	1	6.7	6.7	33.3
.34	2	13.3	13.3	46.7
.35	4	26.7	26.7	73.3
.36	1	6.7	6.7	80.0
.37	1	6.7	6.7	86.7
.39	2	13.3	13.3	100.0
Total	15	100.0	100.0	

```
.30  |\\\\\\\\\\ 1
.31  |\\\\\\\\\\ 1
.32  |\\\\\\\\\\\\\\\\\\\\ 2
.33  |\\\\\\\\\\ 1
.34  |\\\\\\\\\\\\\\\\\\\\ 2
.35  |\\\\\\\\\\\\\\\\\\\\\\\\\\\\\\\\\\\\\\\\ 4
.36  |\\\\\\\\\\ 1
.37  |\\\\\\\\\\ 1
.39  |\\\\\\\\\\\\\\\\\\\\ 2
     |
     |_____
     0         1         2         3         4         5
```

Mean	.345	Std err	.007	Median	.350
Mode	.350	Std dev	.026	Variance	6.9810E-04
Kurtosis	-.358	S E Kurt	1.121	Skewness	.191
S E Skew	.580	Range	.090	Minimum	.300
Maximum	.390	Sum	5.170		
Valid cases	15	Missing cases	0		

Figure 4.2 Data Analysis: Length of Sampled Cables *before* the Cutter Change.

$$t = \frac{\text{observed difference}}{\text{unexplained variability}}$$

where, conceptually, we think this is equivalent to:

$$t = \frac{\text{treatment effect}}{\text{experimental error}}$$

and where (the last implication), we believe that we have actually tested:

$$t = \frac{\text{difference observed due to the change in cutter design}}{\text{experimental error}}$$

Another way of considering the numerator is to think of $\overline{X}_1$ as an estimator of mean deviation from target length before the cutter change, and $\overline{X}_2$ as an esti-

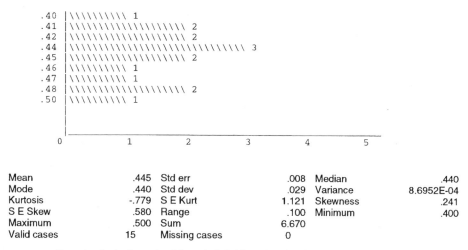

Figure 4.3 Data Analysis: Length of Sampled Cables *after* the Cutter Change.

mator of mean deviation from target length after the cutter change, with an underlying assumption that only the cutter change could account for or explain any difference in a change in mean length deviation from target. Consider the nature of the design. This conclusion is simply not defensible. How do we know that something else did not change? Perhaps the measurement process went out of calibration from the "before" period to the "after" period? What about incoming material? Did the operators on the line change? This design does not allow for a true comparative analysis, because it does not allow us to relate the change in the dependent variable to the treatment effect, and *only* the treatment effect. In fact, the cutter change may have had no effect at all, but an extraneous variable may have been confounded with the treatment. For example, suppose that the "old cutter" data were drawn from the cable universe when the day shift was working. Suppose further that the "new cutter" data were drawn from the cable universe during the night shift. It is possible that the difference we are observing is due to shift differences and not cutter design.

SPSS for MS WINDOWS Release 6.1

t-tests for Independent Samples of DESIGN

Variable	Number of Cases	Mean	SD	SE of Mean
LENGTH				
Old Design	15	.3447	.026	.007
New Design	15	.4447	.029	.008

Mean Difference = -.1000

Levene's Test for Equality of Variances: F= .309 P= .583

	t-test for Equality of Means				95%
Variances	t-value	df	2-Tail Sig	SE of Diff	CI for Diff
Equal	-9.78	28	.000	.010	(-.121, -.079)
Unequal	-9.78	27.67	.000	.010	(-.121, -.079)

Figure 4.4 *t*-Test to Compare Cable Length for Two Cutter Designs.

Some individuals might contend that all of these conditions could be listed, and then controlled. Unfortunately, this would apply to only those independent, or extraneous variables, of which we are aware. There would be no protection inherent to this design that would allow us to guard against the confounding, or intermingling, of extraneous variables that we did not know existed. These conditions require us to classify the one-group pretest-posttest design as a nonexperimental (or preexperimental, or quasi-experimental, or pseudoexperimental) research design.

The static-group comparison design. The static-group comparison design is a method by which a researcher might attempt to provide a second group for comparative purposes, thereby including one of the basic elements of an experimental design. This design may be represented by the form:

$$X \quad O$$
$$\overline{}$$
$$O$$

In this case, the appearance of a second group (row) represents the use of a control group. Many individuals believe that a control group is a group that, quite literally, receives no treatment at all, that is, the absence of a treatment effect. In reality, this is often unnecessary, and in industry, unlikely. The literature, in fact, is replete with examples of research conducted where a "no treatment" control group was inappropriately employed. Consider the classic example of the 1931 Lanarkshire Milk Experiment (Sax, 1979). The purpose of the study was to determine whether students drinking raw milk and students drinking pasteurized milk reflected different gains in height and weight over a 4-month period. To accomplish this purpose, three groups of thousands of students were formed. One experimental group received three-quarters of a pint of raw milk in school on a daily basis. The second experimental group received three-quarters of a pint of pasteurized milk each day in school. A third group, the "control" group consisted of a few thousand students from which milk was withheld, that is, they received "no treatment." In fact, the third group was unnecessary to answer the research questions associated with the purpose of the study.

In industry, the greatest likelihood is that our "control" groups will consist of either (1) current or existing conditions, or (2) placebo treatments. Let us return to a discussion of the inadequacy of the static-group comparison design in providing us with the capability to make causal inferences.

Carefully review the structure of the design previously presented. Note that there is a dashed line between the two groups. This is because the design does not appear as:

$$R \quad X \quad O$$

$$R \quad \quad O$$

In other words, the two groups of experimental units in the static-group comparison have not been randomly selected and assigned to the treatment. Further, there is no pretest present. As a result, we have no method available to us, as researchers, to defend the assertion that the two groups were equivalent prior to the treatment. Therefore, posttreatment (posttest) differences from one group to another attributed to the treatment may, in fact, have been due to the treatment. The noted difference may also, in fact, have been attributable to a difference in the groups that existed before the treatment was even applied. As a result, the initial differences in the two groups are inextricably confounded with the treatment effect. The dashed line between the two groups signifies the nonequivalent nature of the two groups and indicates to the researcher or manager that this design does not represent a true experiment. We should not, therefore, expect causal statements or conclusions generated from this class of research design on a single iteration.

An overview of the threats to internal validity

We have reviewed a number of specific incidents or effects that could have become confounded with the treatment in our sample experiment. The topic we are actually discussing is the general subject referred to as threats to the *internal validity* of designs. When we use the term *internal validity* on a broad scale, we are referring to the technical soundness of the study, that is, whether it is technically correct. When we speak of internal validity on a narrower basis, as in the current discussion, we are referring to the capability of a particular design to prevent extraneous variables from becoming confounded with the treatment effects. By creating an experimental design that possesses internal validity, the researcher is able to assess the relationship between the treatments and dependent variables in a clear and unambiguous fashion (Sax, 1979).

Campbell and Stanley (1963) published a significant amount of material related to this topic. They were able to identify eight factors or conditions that would jeopardize the internal validity of a research design allowing for the confounding of extraneous variables with the treatment and, therefore, rendering the conclusions (not the statistical results) suspect at best and incorrect at worst. Table 4.1 provides a description and illustration of each of these factors or threats.

Let us compare the one-group pretest-posttest design against this list of potential threats. You will recall that the form of this design was:

$$O \quad X \quad O$$

Self-Review Activity 4.1 Against which of the reviewed threats to internal validity would this design most likely fail to protect us? Describe how and why.

If we refer to Table 4.1, we note that the major threats to internal validity posed by the use of the static-group comparison design would be

- Selection biases
- Mortality
- Interaction of selection and maturation

What makes an experimental design out of a preexperimental design? Any valid experiment will:

1. Utilize randomization in sampling, assignment, and order
2. Allow for the assertion of comparative group equivalence
3. Use an experimental design structure, with appropriate tools and methods which when combined with items 1 and 2, will protect us from the major threats to internal validity

Three standard experimental designs which fulfill these requirements are the pretest-posttest control group design, the posttest-only control group design, and the Solomon four-group design. Each of these three designs protects us against the major threats to internal validity.

TABLE 4.1 An Overview of the Major Threats to Internet Validity.

Jeopardizing Factor	Description	Example
History	Specific events, occurring between two sets of measurements or observations that: • Affect the dependent variable, • Are **external** to the treatment, and • Become confounded with the treatment. Historical effects, therefore, have the potential to manifest as changes in the dependent variable, attributed to the treatment effect, where the dependent variable may not have been affected by the treatment at all.	In a One Group Pretest-Posttest Design related to a wave soldering process, a number of boards are run through the process in the morning. The criterion measure is noted, and a treatment is initiated (e.g., a change in targets for set points on speed and temperature). The process is sampled in the afternoon, and a group of new boards are measured. A statistically significant difference, and in fact, an improvement, is noted between the first and second group of experimental units. The new target values are standardized into the process. The next morning, running at the new targets, the boards are unchanged from past levels of quality based upon the criterion measures previously employed. The difference originally attributed to the treatment is subsequently found to have been actually caused by a change in ambient humidity level from the morning to the afternoon, that had been confounded with the treatment.
Selection (Biases)	Inappropriate randomization techniques are employed (sampling, assignment, order), resulting in differences between comparative groups attributed to the treatment effect. Experiments require that groups of experimental units are selected and treatments assigned such that their comparability may be assured.	A plant is testing the effects of chrome versus steel bodymaker tooling on axial load values. Five coils are run from inventory on bodymakers with chrome tooling, five coils are run from inventory on bodymakers with steel tooling. The axial load values for cans made from chrome tooling have a significantly higher average and lower variability than those cans produced with steel tooling. The firm purchases new chrome tooling and installs it throughout the system, with the result of . . . no improvement. The five coils run on chrome tooling were from one manufacturer. The five coils run on steel tooling were from another manufacturer.
Selection-Maturation Interaction	A combinatorial effect of a selection bias in obtaining experimental units for use in the experiment, and the changes taking place within the units due to a maturation effect. In other words, following a nonrandom sampling, assignment, or ordering effort, some experimental units change • In a different way than units in a comparative group, or • In the same way as a comparative group but at a different rate Either condition becomes ambiguously confounded with the treatment.	A transportation manager wishes to test a new type of routing system for delivery trucks against the current system. The criterion measure in this experiment is the conformance to customer promised date (CPD). Selecting a group of drivers from two geographic delivery regions known to have equivalent customer delivery requirements, the manager assigns one region's drivers to use the current system; the other region's drivers are assigned the new system, that is quite unique and different than the method that has been used in the past. Regardless of this uniqueness, the new system is shown to be superior to the existing system. When implemented across the entire organization, it is a catastrophic event, and fails miserably. As it turns out, the new system requires significantly more attention and effort to use than the old system. As the work day progresses, **fatigue** on the part of the driver can set in, and create confusion. In the initial experiment, the region drivers selected to test the new system happened, by chance, to be quite younger (22 to 28 years old) than the rest of the drivers in the delivery system (47 to 55 years old). As the fatigue factor set in with the latter group, the requirements of the system had a greater impact on the drivers, causing customers at the end of the route to experience significant delays.

TABLE 4.1 An Overview of the Major Threats to Internet Validity. (*Continued*)

Jeopardizing Factor	Description	Example
Maturation	Effects occurring **within** (internal to) the subjects or experimental units that: • Are irrelevant as related to the treatment, and • Are associated with the passage of time, and that are • Not specific to particular events. (e.g., fatigue, run-in, etc.)	A Natural Work Group (NWG) is testing two alternate methods for a new Standard Operating Procedure (SOP) for set up and production ramp up of a 6-ton punch press. The dependent variable associated with the treatment is the **time** required to arrive at a stable state of production, as measured from a "cold start" to full speed production. The first SOP is tested and production is achieved in 3 hours and 55 minutes. After a 30 minute lunch, the second SOP is tested, and takes 4 hours and 25 minutes. Obviously, the first time period is better than the second for the first iteration. The exercise is repeated with the same crew for two weeks using the identical test pattern (SOP 1 vs. SOP 2). After this period, SOP 1 is shown to have a lower average time requirement that SOP 2, and is implemented. Remarkably, a subsequent benchmarking effort shows that **SOP 2** is the industry standard. Just for "fun," the crew tries SOP 2 (the rejected procedure) one morning, and achieves a start up time of 2 hours and 37 minutes.
Testing	Effects which show up on a second observation or test, that are affected by an initial or prior test of the experimental unit.	In order to test the effects of a new mounting method for wiper motors, 25 units are randomly selected and tested for noise under load on a test stand with a current mount. The noise levels are recorded, and then each motor is retested for noise level while supported in a mount of a new design. The second set of noise level values are shown to be significantly less than the first group. Even if the second group of tests had been conducted with the "old" mounting method, the noise levels would have been lower. New motors stressed and tested for noise under load become quieter until the run-in period is completed and noise level "flattens out," or stabilizes.
Instrumentation	Effects confounded with the treatment that are specifically related to changes or shifts in measurement processes. These effects originate in two domains: • Bias, the error associated with accuracy and/or • Measurement error, the error associated with precision. For discrete data, this would be expressed as changes in the measurement system's: • Validity and/or • Reliability	A company has justified the purchase of a new printer/coater on the basis of the new equipment's potential capability to improve print quality. The criterion measure of primary interest, based upon the customer quality assurance system, is the visually assessed "vividness" or "intensity" of the colors red and blue. In the past, these measures have been assessed on an ongoing basis by Monte (**exclusively**), the head printer, using a standard scale located in the printing department. The new equipment is purchased and installed, and tested on the basis of the criteria of interest. The test is conducted with the agreement of the equipment vendor that if the color intensity has not been improved, we will not pay for the equipment. During the interim period, Monte quit. His "right hand man," Melissa, has been promoted to the head printer position. She assesses the new product off the tested equipment, and the statistical facilitator shows that there has been a significant improvement in the criteria of interest. The equipment is paid for, and production is initiated. After three months, the external customer has noted no improvement in product quality.

Defending the Design

TABLE 4.1 An Overview of the Major Threats to Internet Validity. (*Continued*)

Jeopardizing Factor	Description	Example
Statistical Regression	A result of the nonrandom selection of subjects, experimental units, or groups on the basis of their extreme scores or characteristics. Scores of such units or groups tend to regress toward the mean on repeated measures through time. This change could be attributed to the treatment, when in fact the shift would have occurred in the absence of the effect. Since scores at the extreme of a distribution are more subject to chance effects and factors than scores at the central portion of a distribution, the units or subjects selected for their extreme characteristics are more likely to have true values closer to the mean. These true values, when made robust against the chance, one-time effects associated with repeated or more frequent measures, are likely to become less extreme (as a group).	A new secondary heat treat process is being tested to determine whether it would have the ability to reduce the variability of hardness in our current production system (we manufacture saw chain cutters). A series of lots for our process has been tested and segregated based on their extremely high and low hardness values. Sample units from the "HIGH" lots and "SOFT" lots are processed through the secondary process (the treatment), and retested. The variability between the lots, after testing 30 units in each processed lot, is shown statistically to have been reduced. The new process is implemented. Tracking the production process on the standard control charts used in the past, however, shows that no significant reduction in the variability of the lots has been achieved.
Experimental Mortality	Loss of experimental units or subjects during the conduct of the experiment.	An experiment is conducted to determine whether a new laboratory test may be used to predict an equivalent result when the same product is tested in the field. A random sample of product in "matched pairs" (i.e., homogeneous blocks) is selected, and one-half of each pair is sent to either the lab or field. This testing is extremely expensive, so the minimum number of experimental units are used. In the process of the experiment, some of the field test units are lost or misplaced. Since the study was so expensive, the remaining data were assessed anyway, and the laboratory test was shown to be a predictor of field results. A new product has been designed, tested, and introduced solely on the basis of the newly incorporated laboratory test. The field results are now being reported, and the product is not performing as the laboratory test suggested it would. A second experiment on the predictive capability of the laboratory test is quite revealing. As it turns out, by chance and chance alone, the original lost units probably were key and, had the data have been available, would have changed both the nature of the prediction as well as the confidence of personnel in the accuracy of the prediction.

Designs which protect against major threats to internal validity

Pretest-posttest control group design. The pretest-posttest control group design appears as follows:

$$R \quad O \quad X \quad O$$
$$R \quad O \quad \quad O$$

Let us briefly review an experiment conducted using this design, as an example of how it might be applied, allowing us to make causal inferences.

Description of the problem. A transportation group manager wishes to determine whether maintenance costs vary for fleet cars using unleaded medium-grade gasoline (our current fuel of choice by policy) or a new gasohol blend. Maintenance costs in this case include fuel, oil, and all required scheduled and nonscheduled maintenance costs. The current fleet consists of 1250 vehicles, each driven by the same employee (i.e., no employee ever drives someone else's vehicle). The manager wishes to conduct a valid study that will allow her to determine which is the "superior" fuel blend, based upon the stated criterion measure.

Description of the study. From the fleet pool in St. Louis, 250 vehicles are randomly selected and further subdivided (randomly, of course) into two groups. The two groups are tracked through an appropriate period of time, with the necessary data collected in order to show that the two groups are equivalent on the basis of the criterion measure. The manager then randomly selects one of the two groups to start using gasohol; no change is made to the policy employed by the second group. The two groups are tracked for a period of 12 months, and the resultant data are properly analyzed.

The statistical analysis of the data is utilized to determine whether the entire fleet should change over to gasohol (which has a higher price, but perhaps a lower cost to use when maintenance costs are considered) or stay with medium-grade gasoline.

Incidentally, note that we did not include a control group in the sense of a "no treatment" group of vehicles. We know that maintenance costs are lower for parked cars.

Self-Review Activity 4.2 Describe a potential threat to internal validity for a research study of this type, based upon each of the following sources: (1) history, (2) maturation, (3) selection, and (4) mortality. Describe how this experimental design would protect the researcher (manager) against extraneous variables becoming confounded with the treatment.

Posttest-only control group design. The posttest-only control group design has the form:

$$R \quad X \quad O$$

$$R \quad \quad O$$

While similar to the static-group comparison, the groups are considered equivalent. Why would we not use a pretest? First, we might have a situation where we utilize a Class I or Class II destructive test. (A Class I destructive test is where the samples measured are destroyed by the measurement system, but samples may be taken from homogeneous lots or subgroups. A Class II destructive test is where the samples are destroyed by the measurement system but only heterogeneous lots or batches are available for sampling.) In these cases, the specimen or experimental unit will be destroyed. Posttesting would present some definite problems. Second, we might be worried about the external validity, or the ability to generalize our results to the research and/or target population. Basically, we might be concerned as to whether pretested experimental units respond to the treatment effect in the same way that experimental units not pretested will respond. Since our universe consists of units not pretested, this might be a valid concern.*

A second interesting aspect of this design is the query: On what basis, in the absence of a pretest, may we defend the assumption that we started with equivalent groups? The answer to this question is, of course, that in this case (unlike the static-group comparison), we employed randomization. Applied correctly,† randomization serves as the process by which groups can be equated, provided the samples or groups are large enough to justify the assumption that all extraneous, nonmanipulable variables have been equalized between the test groups. The issue of "How large is large?" will be discussed in Chap. 6.

The Solomon four-group design. A third true experimental design offering protection against potential threats to the internal validity of our design is the Solomon four-group design. This design, which combines the strengths of the pretest-posttest control group design and the posttest-only control group design while generally eliminating their weaknesses, has the form:

$$R \quad O \quad X \quad O$$
$$R \quad O \quad \quad O$$
$$R \quad \quad X \quad O$$
$$R \quad \quad \quad O$$

Self-Review Activity 4.3 Describe how the Solomon four-group design is stronger than the previous two designs.

Figure 4.5 summarizes the designs we have discussed, and details an assessment by Campbell and Stanley (1963) of the protection provided by each design against the major threats to internal validity reviewed.

With the presentation of this matrix, we have completed our discussion of the experimental design outcome in which the treatment is not related to/does not affect the dependent variable but the experiment causes the researcher to

---

*Threats to external validity will be presented after the discussion of Fig. 4.1 is completed.
†This entire discussion, as well as a significant portion of the subsequent discussions of threats to internal validity, are based upon the premise that researcher errors are absent. We are distinguishing here between systematic errors made due to poor or inadequate design choices versus systematic errors due to mistakes (e.g., incorrect statistical tests selected, incorrect recording of data, misunderstanding of test instructions).

	Sources of Internal Invalidity							
	History	Maturation	Testing	Instrumentation	Regression	Selection	Mortality	Interaction of Selection and Maturation
Pre-Experimental Designs:								
1. One-Shot Case Study X O	−	−				−	−	
2. One-Group Pretest-Posttest Design O X O	−	−	−	−	?	+	+	−

3. Static-Group Comparison
  X   O
  ‑ ‑ ‑ ‑ ‑
       O
	+	?	+	+	+	−	−	−

True Experimental Designs:								

4. Pretest-Posttest Control Group Design
  R   O   X   O
  R   O   O
	+	+	+	+	+	+	+	+

5. Solomon Four-Group Design
  R   O   X   O
  R   O   O
  R   X   O
  R   O
	+	+	+	+	+	+	+	+

6. Posttest-Only Control Group Design
  R   X   O
  R   O
	+	+	+	+	+	+	+	+

Legend: +   Controls for effects
          −   Does not control for effects
          ?   May control for effects under certain conditions
          ‑‑‑‑‑‑ Nonequivalent groups

Campbell, Donald T. and Julian C. Stanley, *Experimental and Quasi-Experimental Designs for Research*. Copyright © 1963 by Houghton Mifflin Company. Used with permission.

Figure 4.5 Sources of Invalidity for Some Pre-Experimental and True Experimental Designs.

infer that a relationship/effect does exist. This outcome was presented in Fig. 4.1.

Type II error: The treatment is related to/does affect the dependent variable, but the experiment causes the researcher to infer that no relationship/no effect exists.

The other incorrect conclusion that could be drawn is that the treatment is related to/does affect the dependent variable, but the experiment causes the researcher to infer that no relationship/no effect exists. This particular outcome, called a Type II error is, in many ways, worse than the first outcome we reviewed. In the first instance, we would at least eventually discover that the original experiment was flawed, as we found ourselves unable to replicate the results in our universe. In this case, however, we would probably be facing the potential of a lost opportunity, in that the failure to recognize a potential "solution" would probably cause us to abandon the approach, possibly forever. This design failure could occur due to any of the following conditions:

- An extraneous, manipulable independent variable is confounded in the denominator
- The selected experimental design lacks sensitivity
- Extraneous variables are confounded in the numerator or other threats to internal validity contribute to the previous design failure

We will now explore each of these potential hazards.

An extraneous, manipulable independent variable becomes confounded in the denominator. In order to understand the condition where an extraneous, manipulable independent variable becomes confounded in the denominator, we will combine our previous discussion on statistical inference with the topic of experimental design in the following example. A company manufactures spar tabs which are components that are used in the production of airplane wings. A critical product characteristic of spar tabs is the ability to withstand stress in flight which is measured by a test of cycles to failure. Obviously, the company wants to maximize cycles to failure, balanced with production costs. Assume that we have generated the following statement of the problem:

> The purpose of this study is to determine whether the cycles-to-failure level of spar tabs produced from ingots processed from our current alloy is different from that for spar tabs produced from ingots processed with a new alloy recommended by our corporate research and development group.

The (primary) research hypothesis posed for this study was

> Spar tabs produced with the new alloy will have a higher average cycle-to-failure level on the 45° bend test than spar tabs manufactured using our current alloy.

The experimental design selected for this study was a posttest-only control group design:

$$R \quad X \quad O$$

$$R \quad X_C \quad O$$

where X_C indicates the presence of a control, or current condition, comparative group. No classical control group will be employed with this experiment; we already understand that failing to heat-treat material leads to lower cycles-to-failure values.

The appropriate sample size required for this experiment was calculated based on an effect size of 10 units (on the average). The quality improvement team (QIT) leader informed the team sponsor that 16 ingots at each level of the treatment variable had to be processed through the heat-treat furnace, for a total of 32 ingots. Given that all previous research has shown that each ingot is relatively homogeneous, only one sampled spar tab would have to be tested from each ingot. In other words, we will obtain one measure per ingot.

The sponsor spoke to the plant manager, and after a review of the design proposal, the experiment was approved.

It was obvious to all of the personnel involved in the study that this experiment was going to take some time. The single furnace can hold only two ingots per production load. The ingots to be processed with the new alloy were randomly assigned to blocks, or group numbers, 1 through 16. The ingots manufactured with our current alloy were randomly assigned block numbers 1 through 16 as well. Randomly selecting one ingot from each treatment level, 16 blocks of two ingots were created, with each block consisting of a proposed alloy ingot and a current alloy ingot.

Each pair of ingots were processed through the heat-treat furnace in a random order. The ingots were also randomly assigned to their respective position within the furnace itself. In 16 of the 32 cases, the current alloy ingot was placed on the right-hand side (R) of the furnace, and the proposed alloy ingot was positioned to the left-hand side (L). In the other 16 cases, the positions were reversed. The order of the positioning was randomized through the processing order (e.g., L/R, R/L, R/L, L/R R/L, etc.).

The ingots were processed (heat-treated) according to our current standard operating procedure (SOP). All other conditions were held to our current SOP set points.

Each ingot was randomly sampled at the end of the line, and we obtained our 32 spar tabs, 16 from the new alloy, 16 from the old alloy. The tabs were randomly assigned to 16 blocks of level pairs, and randomly tested in our lab, with the order of testing alternated (i.e., current/new, new/current) through the test period.

The statistical hypotheses tested appeared as

$$H_0: \mu_{NEW} = \mu_{CURRENT}$$

$$H_1: \mu_{NEW} \neq \mu_{CURRENT}$$

The data collected as related to time (how we measure cycles) appeared as shown:

Alloy (Treatment)

New (proposed)	Current
246	317
296	238
312	350
243	314
378	359
325	284
251	286
381	229
372	287
310	349
249	362
377	315
300	231
327	279
322	313
341	235

The analysis of the data was generated and presented to the team sponsor as illustrated in Fig. 4.6.

As we review these data, and their associated statistical analysis, we may observe that while the observed mean difference exceeded the effect size originally targeted (10), the calculated probability that the observed difference is due to sampling error, or chance, is 30 percent (refer to the p value calculated on the t-test in Fig. 4.7). As a result, it is possible that we would now move on, believing that the treatment was ineffective.

Unfortunately, this would be a tragic mistake. The error we have made basically corresponds to the mistaken belief held by many industrial personnel who design experiments that you "study what you're interested in, and randomize everything else." This approach, while definitively avoiding the effect of confounding known manipulable variables with the treatment, results in artificially inflating the denominator, the experimental error estimate. The denominator of the test statistic, a representation (in this case) of the experimental error in the model, is intended to represent unknown and uncontrollable effects and extraneous variables. We have, in effect, violated this assumption. What we have calculated in the denominator of the t-test statistic is actually two effects: (1) experimental error and (2) the variability in the 32 observations explained by run-to-run differences. That is, the error term consists of the variability in the failure times explained not by the treatment, not by unknown and uncontrollable variables, but by the differences in heat-treat load. We may illustrate this effect in a number of ways.

First, review the variances associated with the two levels of alloys (in Fig. 4.6) that are pooled and used for the denominator of the test statistic, and the estimate of experimental error.

SPSS for MS WINDOWS Release 6.1

Number of valid observations (listwise) =   16.00

Variable CURRENT   Current Alloy

Mean	296.750	S.E. Mean	11.523
Std Dev	46.093	Variance	2124.600
Kurtosis	-1.161	S.E. Kurt	1.091
Skewness	-.161	S.E. Skew	.564
Range	133.000	Minimum	229.00
Maximum	362.00		

Valid observations -   16   Missing observations -   0

Variable PROPOSED   Proposed Alloy

Mean	314.375	S.E. Mean	12.145
Std Dev	48.578	Variance	2359.850
Kurtosis	-1.107	S.E. Kurt	1.091
Skewness	-.142	S.E. Skew	.564
Range	138.000	Minimum	243.00
Maximum	381.00		

Valid observations -   16   Missing observations -   0

- - Description of Subpopulations - -

Summaries of   CFL   Cycle to Failure Level
By levels of   ALLOY

Variable	Value	Label	Sum	Mean	Std Dev	Variance	Cases
For Entire Population			9778.00	305.5625	47.4348	2250.0605	32
ALLOY	1.00	Proposed	5030.00	314.3750	48.5783	2359.8500	16
ALLOY	2.00	Current	4748.00	296.7500	46.0934	2124.6000	16

Total Cases = 32
Missing Cases = 0

Figure 4.6 Descriptive Statistics for Cycles-to-Failure Level for Proposed and Current Alloys.

It should not surprise you that the difference in means was attributed to chance. Given the size of the variances within each level, the outcome is not surprising. But, we also recognize that if there were significant differences that could be explained by between load effects, this variation would have been confounded within the treatment levels. Specifically, suppose we have what we think is a single group of 10 specimens, providing the following data set:

t-tests for Independent Samples of ALLOY

Variable	Number of Cases	Mean	SD	SE of Mean
CFL Cycle to Failure Level				
Proposed	16	314.3750	48.578	12.145
Current	16	296.7500	46.093	11.523

Mean Difference = 17.6250

Levene's Test for Equality of Variances: F= .002 P= .968

	t-test for Equality of Means				95%
Variances	t-value	df	2-Tail Sig	SE of Diff	CI for Diff
Equal	1.05	30	.301	16.742	(-16.566, 51.816)
Unequal	1.05	29.92	.301	16.742	(-16.570, 51.820)

Figure 4.7 t-Test to Compare the Mean Cycles-to-Failure Level for Proposed Versus Current Alloy.

$$23, 45, 22, 45, 24, 46, 43, 47, 22, 21$$

This gives us a calculated variance (think of this value in terms of within-group, or unexplained, variation) of 145.96. Suppose, however, that the 10 measures actually were produced by randomly assigning the 10 values to two test stands and that we should really represent the data as

Stand (block) 1   23
          22
          24     $s^2 = 1.30$
          22
          21

Stand (block) 2   45
          45
          46     $s^2 = 2.19$
          43
          47

Then, the more accurate estimate of experimental error would be not 145.96, but the average within-block variance of 1.745 (s^2). What is the source of variability accounting for the difference between the two values? Obviously, it is the difference in test stands. Of course, if the data had appeared as

82   Chapter Four

Stand (block) 1   23
                   46
                   24    $s^2 = 152.5$
                   45
                   22

Stand (block) 2   22
                   21
                   45    $s^2 = 167.8$
                   43
                   47

then we would recognize that the test stand effect was inconsequential as a source of explained, or explainable, variability. If we randomize our study across test stands, and the resultant data appears as shown in the second case, there is no problem (this does not imply that it was a good idea, it simply implies that we have not artificially inflated our error term). If, on the other hand, the data should appear as shown in the first case, we have ignored a source of variation from a known manipulable extraneous independent variable, resulting in an artificial inflation of our experimental error term which significantly compromises the sensitivity of our experiment.

Of course, the question remaining is to determine into which category our

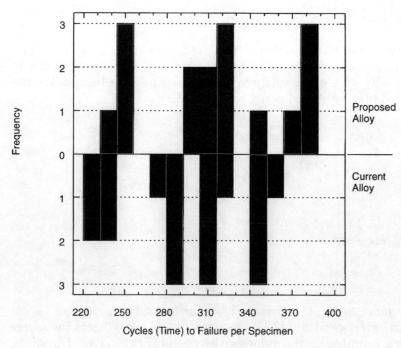

Figure 4.8 Comparative Histograms for Time to Failure for Tested Alloys.

```
Summaries of   TIME
By levels of   LOAD

Variable           Value      Label       Mean      Variance      Cases

For Entire Population                   305.2500    2216.7742      32

LOAD               1.00                  236.0000     98.0000       2
LOAD               2.00                  238.5000    112.5000       2
LOAD               3.00                  242.0000     98.0000       2
LOAD               4.00                  244.5000     84.5000       2
LOAD               5.00                  287.5000    144.5000       2
LOAD               6.00                  292.0000    128.0000       2
LOAD               7.00                  298.0000    288.0000       2
LOAD               8.00                  299.5000    312.5000       2
LOAD               9.00                  317.5000     40.5000       2
LOAD              10.00                  319.5000     60.5000       2
LOAD              11.00                  321.0000     72.0000       2
LOAD              12.00                  329.0000    288.0000       2
LOAD              13.00                  360.0000    264.5000       2
LOAD              14.00                  363.5000    364.5000       2
LOAD              15.00                  365.0000    338.0000       2
LOAD              16.00                  370.0000    242.0000       2
```

Total Cases = 32

Figure 4.9 Descriptive Statistics for Cycles-to-Failure Levels by Production Load.

heat-treat production lot effect falls. Review Fig. 4.8. What is your interpretation based upon a visual analysis of the data presented?

Another method of "breaking out" the data would be to generate descriptive statistics for the failure data by production load as illustrated in Fig. 4.9.

Review the variance within lot values in Fig. 4.9. Compare the variation, which is due to the variability between ingots within each load, to the variance due to load variability within each alloy (see Fig. 4.6).

As shown by both Fig. 4.10 and an examination of the descriptive statistics (see Fig. 4.9), a significant amount of the variation in the model was due to differences in failure rates due to load. By statistically accounting for this effect in a correct fashion, we can, in fact, "remove" this estimate of variation from our hypothesis test as related to the test for our treatment effect.

Recall that our previous t value and p value (see Fig. 4.7) were calculated as

$$t = 1.05 \quad p = 0.301$$

Correctly assessing alloy (treatment) differences in failure rate within blocks (loads) rather than across blocks, we are capable of determining the true alloy effect (as presented in Fig. 4.11).

We can easily see why we would now reject the null hypothesis that we previously almost accepted. When the first analysis was conducted, and our load effect was confounded in the denominator, we were actually comparing means

Time to Failures for Tested Specimens

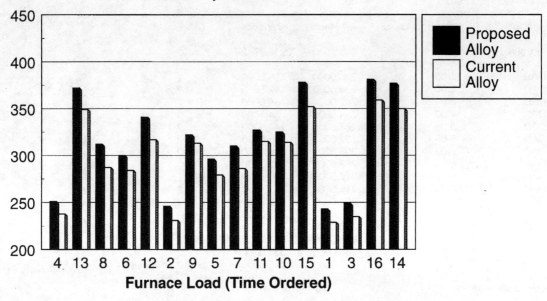

Figure 4.10 Analysis of Time to Failure for Proposed Versus Current Alloys by Furnace Load.

Paired samples t-test: PROPOSED
CURRENT

Variable	Number of Cases	Mean	Standard Deviation	Standard Error
PROPOSED	16	314.3750	48.578	12.145
CURRENT	16	296.1250	45.209	11.302

(Difference) Mean	Standard Deviation	Standard Error	Correlation	2-Tail Probability	t Value	Degrees of Freedom	2-Tail Probability
18.2500	6.017	1.504	.994	.000	12.13	15	.000

Figure 4.11 Paired Samples t-Test to Compare Effect of Current Versus Proposed Alloy on Cycle Time to Failure Levels.

with experimental error levels represented by the two confidence intervals illustrated in Fig. 4.12.

Eliminating the variability actually due to the production load (block) variability from the error term (denominator), we are now comparing means with error terms represented by the confidence intervals illustrated in Fig. 4.13.

The moral of this story is not, you should note, that statistical methods will always be available to "save us." If load variability should have been studied, our original sample size calculation was probably wholly inadequate. Further, if we did not have an engineering log (discussed in Chap. 8) for the experiment

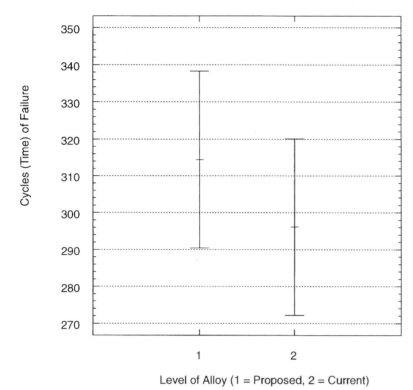

Figure 4.12 95 Percent Confidence Intervals for Alloy Level Means When Load Effect is Confounded with the Error Term.

allowing us to identify load numbers by ingot, we would not have been able to even account for this extraneous variable.

The purpose of presenting this example was to show the effects of randomizing across known and manipulable independent variables, and the effect of that approach on the sensitivity of an experiment. All experimental designs must be carefully planned, designed, and controlled. Randomization, and randomization alone, is not an appropriate method for managing known independent yet extraneous variables. The error term is referred to as *unexplained variation* and should be thought of as reserved for this effect. Any explainable variation due to a known and manipulable independent variable that is permitted to confound this term through the abrogation of design methodology, and a dependence on randomization, violates the premises of sound experimental design.

Incidentally, one of the issues that would be of interest relates to whether the failure results were related to the order in which they were processed. In fact, again, given the presence of an engineering log, we can show that there was no such relationship. Figures 4.14 and 4.15 present the scatter plots between the order of processing (an extraneous variable catchall) and the time

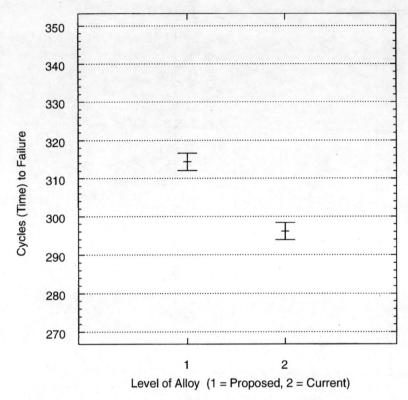

Figure 4.13 95 Percent Confidence Intervals for Alloy Level Means when Production Load is Blocked.

to failure recorded for the specimen tested for each alloy (proposed and current).

The correlation coefficients (a measure of the relationship between factors) for these data appear as shown below (more information on this topic will be presented in a subsequent chapter).

Spearman Rank Correlations

	Time order*
Proposed alloy	.2324/(16)/.3682
Current alloy	.2324/(16)/.3682

*Coefficient/(sample size)/significance level.

Significance levels of .3682 indicate that there is no significant relationship between the order of processing and the time to failure for either the proposed or current alloy.

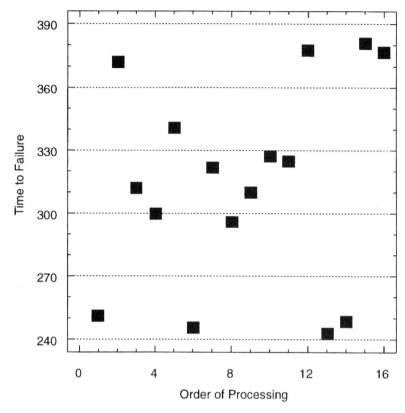

Figure 4.14 Plot of Time to Failure Values for Proposed Alloy by Heat-Treat Process Order.

The selected experimental design lacks sensitivity. A second opportunity for concluding that a treatment has no effect, when in reality, it does (refer again to Fig. 4.1), could result from the relationship between the treatment effect and the level of experimental error—specifically, from an experimental design that lacks sensitivity.

The term *sensitivity* refers to the precision or power of an experiment to detect actual treatment effects. The last example should allow you to note that a given difference in the numerator may or may not allow us to attribute the difference as statistically significant, and that this result relates directly to the size of the error term. The sensitivity of an experimental design may be increased or controlled in four ways:

1. Increasing the sample size
2. Testing the full strength of the treatment
3. Reducing the variability of chance and random effects
4. Utilizing experimental designs that maximize treatment sensitivity and minimize experimental error effects

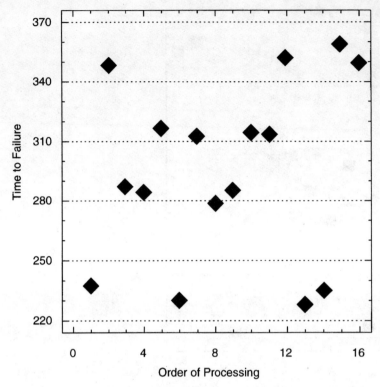

Figure 4.15 Plot of Time to Failure Values for Current Alloy by Heat-Treat Process Order.

We will explore each approach.

Increasing sample size. Increasing sample size within each level or treatment combination tested is a highly effective method for increasing the sensitivity of an experiment. Sample size is a function of the effect size to be detected, the number of levels or treatment combinations to be tested, the inherent variability of the populations involved in the study, the probability of committing Type I and Type II errors that the researcher is willing to accept, and the economic constraints of the study. The methods by which sample sizes may be calculated will be presented in Chap. 6, but the interrelationship among these elements is important to understand as a basic building block of experimental design.

In order to briefly display this condition, examine the data presented in Table 4.2. Five finite populations of size $N = 10,000$ were established, with the parameters shown in the heading of the table. Then, a random sample of size 5 was randomly drawn from each of the populations, and statistically compared with the other samples. For each independent pair of samples, a t value and a p value were calculated for statistical comparison. This procedure was

TABLE 4.2 Summary of Monte Carlo Simulation for Comparative Group Analysis

Sample size = 5:	Group 1 versus Group 2			Group 1 versus Group 3			Group 4 versus Group 5		
	$\mu = 100$, $\sigma = 10$ versus $\mu = 110$, $\sigma = 10$			$\mu = 100$, $\sigma = 10$ versus $\mu = 120$, $\sigma = 10$			$\mu = 100$, $\sigma = 2$ versus $\mu = 110$, $\sigma = 2$		
	X-bar difference	t*	p	X-bar difference	t*	p	X-bar difference	t*	p
	6.91	1.17	.274	18.11	2.04	.076	8.22	6.99	.000
	8.03	1.27	.239	17.76	2.18	.060	8.83	7.70	.000
	13.26	1.28	.236	22.74	2.55	.034	9.60	8.25	.000
	11.40	1.46	.181	17.36	3.30	.011	11.96	8.57	.000
	9.19	1.49	.174	16.52	3.61	.007	11.51	8.66	.000
	7.10	1.62	.144	16.41	3.65	.006	10.02	8.88	.000
	13.03	1.66	.136	30.40	3.70	.006	8.14	11.04	.000
	22.99	3.25	**.012**	23.70	3.90	.005	10.09	11.52	.000
	14.56	3.34	**.010**	20.16	7.25	.000	12.33	12.58	.000
	11.72	4.02	**.004**	25.64	8.20	.000	11.39	12.61	.000
	Mean Difference = 11.8			Mean Difference = 20.9			Mean Difference = 10.21		

* Absolute value for t. Bold p-values indicate rejected hypothesis of equality at $\alpha = .05$

Sample size = 10:	Group 1 versus Group 2			Group 1 versus Group 3			Group 4 versus Group 5		
	$\mu = 100$, $\sigma = 10$ versus $\mu = 110$, $\sigma = 10$			$\mu = 100$, $\sigma = 10$ versus $\mu = 120$, $\sigma = 10$			$\mu = 100$, $\sigma = 2$ versus $\mu = 110$, $\sigma = 2$		
	X-bar difference	t*	p	X-bar difference	t*	p	X-bar difference	t*	p
	4.57	1.29	.214	11.46	2.51	**.022**	10.34	9.59	.000
	6.11	1.63	.121	18.92	3.38	**.003**	8.91	9.64	.000
	8.13	1.93	.069	15.06	3.67	.002	8.74	10.41	.000
	9.51	2.05	.055	13.72	3.91	.001	10.46	11.14	.000
	13.37	2.15	.045	18.73	4.30	.000	10.41	11.43	.000
	9.58	2.18	.043	27.21	4.53	.000	10.46	12.88	.000
	15.33	2.76	.013	21.18	4.67	.000	11.07	13.70	.000
	13.94	3.65	.002	16.89	5.45	.000	10.46	15.29	.000
	19.41	4.74	.000	24.43	6.07	.000	11.51	15.30	.000
	10.91	5.70	.000	29.39	8.71	.000	8.49	18.61	.000
	Mean Difference = 11.1			Mean Difference = 19.70			Mean Difference = 10.08		

* Absolute value for t. Bold p-values indicate rejected hypothesis of equality at $\alpha = .05$

TABLE 4.2 Summary of Monte Carlo Simulation for Comparative Group Analysis (*Continued*)

Sample size = 20:	Group 1 versus Group 2			Group 1 versus Group 3			Group 4 versus Group 5		
	$\mu = 100, \sigma = 10$ versus $\mu = 110, \sigma = 10$			$\mu = 100, \sigma = 10$ versus $\mu = 120, \sigma = 10$			$\mu = 100, \sigma = 2$ versus $\mu = 110, \sigma = 2$		
	X-bar difference	t*	p	X-bar difference	t*	p	X-bar difference	t*	p
	1.91	.69	.494	12.69	4.54	**.000**	9.56	14.65	**.000**
	7.36	2.08	**.044**	17.60	5.30	**.000**	10.99	14.87	**.000**
	7.86	2.58	**.014**	18.44	6.66	**.000**	9.47	15.10	**.000**
	9.42	2.96	**.005**	21.47	6.71	**.000**	9.20	16.02	**.000**
	10.25	3.16	**.003**	22.18	6.80	**.000**	9.43	16.69	**.000**
	9.85	3.28	**.002**	19.28	7.28	**.000**	9.60	17.29	**.000**
	12.46	3.58	**.001**	19.15	7.40	**.000**	10.65	17.38	**.000**
	13.40	3.61	**.001**	23.08	7.58	**.000**	8.42	17.76	**.000**
	11.28	3.69	**.001**	21.32	8.34	**.000**	10.96	18.09	**.000**
	12.51	3.78	**.001**	19.94	10.10	**.000**	10.66	18.44	**.000**
	Mean Difference = 9.63			Mean Difference = 19.5			Mean Difference = 9.89		

* Absolute value for t. Bold p-values indicate rejected hypothesis of equality at $\alpha = .05$

Sample size = 50:	Group 1 versus Group 2			Group 1 versus Group 3			Group 4 versus Group 5		
	$\mu = 100, \sigma = 10$ versus $\mu = 110, \sigma = 10$			$\mu = 100, \sigma = 10$ versus $\mu = 120, \sigma = 10$			$\mu = 100, \sigma = 2$ versus $\mu = 110, \sigma = 2$		
	X-bar difference	t*	p	X-bar difference	t*	p	X-bar difference	t*	p
	4.02	1.97	.052	18.98	8.96	**.000**	9.51	23.00	**.000**
	8.27	4.20	**.000**	19.78	9.38	**.000**	9.55	23.71	**.000**
	9.85	4.74	**.000**	20.23	9.70	**.000**	10.20	24.55	**.000**
	9.59	5.16	**.000**	22.37	9.81	**.000**	10.78	25.02	**.000**
	9.71	5.21	**.000**	17.14	10.26	**.000**	9.60	26.31	**.000**
	10.38	5.34	**.000**	19.02	10.65	**.000**	10.11	26.66	**.000**
	10.38	5.78	**.000**	19.38	10.99	**.000**	9.94	26.71	**.000**
	12.40	5.95	**.000**	19.31	11.56	**.000**	10.23	27.29	**.000**
	10.05	6.05	**.000**	20.07	11.98	**.000**	10.09	27.48	**.000**
	15.32	8.85	**.000**	23.67	13.33	**.000**	10.00	28.76	**.000**
	Mean Difference = 9.997			Mean Difference = 19.995			Mean Difference = 10.001		

* Absolute value for t. Bold p-values indicate rejected hypothesis of equality at $\alpha = .05$

repeated 10 times with samples of size 5. The entire procedure was repeated with the same populations for samples of size 10, 20, and 50.

Examine the results displayed in the figure carefully. What is your assessment of:

- The probability of rejecting the null hypothesis as the sample size increases?
- The probability of rejecting the null hypothesis as the effect size increases?
- The probability of rejecting the null hypothesis as the variability of the population decreases?

Finally, take careful note of the last row on Table 4.2 for each of the four sample sizes. How would you interpret these data, in the context of experimental error and point estimation?

Testing the full strength of the treatment. In conducting any experiment, the researcher wishes to provide the treatment every opportunity to make a difference in the dependent variable studied. Another way of expressing this desire is to note that the levels selected for study should be as far apart as possible, without leaving the operating envelope of the system under investigation. A brief example will clarify this approach.

A major automotive manufacturer wanted to test the effects of burning-in, rather than running-in, electronically tuned radios. Electrical components have a higher probability of failure in the initial stages of the "life" of the part (infant mortality). Manufacturers typically attempt to cause any device failures during the infant mortality time period at the manufacturing facility rather than at the customers' location (to minimize warranty costs and customer dissatisfaction). *Run-in* refers to the process of running the device through the infant mortality time period (also called a *life test*). *Burn-in* refers to accelerating the life test by altering environmental conditions. For example, burning-in by running the product for 8 hours at 140° versus 16 hours at 75° with run-in. There were many aspects to this study. If burn-in did have a significant effect in reducing (external) customer dissatisfaction, then the second issue that had to be answered was related to time. That is, if burn-in proved necessary, how long did the burn-in period have to last in order to obtain its full effect?

In order to answer this question, the treatment was broken into three levels: run-in, burn-in at 4 hours, and burn-in at 8 hours. Without delving into the theories associated with electronics technology, why would we not study the treatment at run-in, burn-in for 30 seconds, and burn-in at 45 seconds? Because this period would not provide burn-in an opportunity to show an effect on final radio failure rate unless sample sizes were astronomically high (refer to the discussion related to Table 4.2). On the other hand, we also would not test (for example) burn-in at an extreme level of 30 days. Management has no intention of burning-in radios for this period due to its inherent cost. Due to the financial implications of such a period, this would be true almost irrespective of the effect of such a duration on final failure rate. As a result, the study would be designed to permit the treatment (burn-in) to display its full effect on the dependent variable, without conducting the analysis outside of the operating envelope to which the results of the study may be generalized.

Reduce the variability of chance and random effects. By reducing the opportunity of chance and random effects to intrude into the experimental space, we limit the expansion of the experimental error term, thereby increasing the sensitivity or power of the experiment. This is generally accomplished through careful design, treating experimental units or subjects in a consistent fashion (e.g., using scripts when collecting data through interviews), and using measurement processes with low inherent error effects (i.e., low levels of measurement error).

Utilizing experimental designs to maximize treatment sensitivity and minimize experimental error effects. The last method available to us for the enhancement of experimental design sensitivity is the utilization of those designs that allow for the inherent minimization of experimental error. The two categories of these designs are generally identified as *planned grouping* or *block* designs, and *repeated measures* designs.

Block designs will be explored and illustrated in Chap. 5. These designs take on a number of forms. A basic and common example of this design category is the matched pairs design. This design was illustrated by our previous discussion corresponding to the spar tab fatigue study. Basically, the matched pairs design is structured to allow us to compare differences within pairs or blocks instead of across the total groups. The result of this effort is an experimental error term reduced by the amount of variability associated with the block-to-block effect.

Repeated measures designs also reduce the experimental error term, in a slightly different way. Rather than use homogeneous pairs or blocks, repeated measures designs employ the tactic of repeatedly measuring the same experimental units or subjects. This process results in a type of before and after, or pretest-posttest, set of measurements. The reduction of experimental error comes from our ability to compare the "before" and "after" results on the same experimental units rather than on a second and different randomly sampled set of units. The anticipated difference encountered between two independent samples due to sampling error alone would be, therefore, eliminated.

As related to Fig. 4.1, these methods allow us to increase the sensitivity (power) of an experimental design, avoiding the possibility that a true treatment effect will be overlooked as a result of a poor experimental design (Type II error).

Confounding or other threats to internal validity. The third source of a design failure which can lead to Type II error (see Fig. 4.1) is confounding or other threats to internal validity. We have already discussed, in depth, the potential threat posed by confounded extraneous variables "masquerading" as treatment effects. In these instances, you will recall, the treatment levels became confounded with the confounding variable, and led us to believe that the observed and statistically significant difference observed was due to the treatment, when in fact, it was the confounded variable causing the effect. This condition was comprehensively discussed in conjunction with the cutter design study and summarized by a description of threats to internal validity presented in Table 4.1.

The issue here is the effect of a confounded variable masking the effect of a treatment which truly does exist. For example, a company manufactures circuit boards. A critical product characteristic is the proper soldering of wires and components so they do not separate from the circuit board, causing it to fail prematurely. Pull force is the force required to pull the wire or component away from the board, but this is a destructive test because the entire circuit board is damaged by the pulling. Therefore, the manufacturer solders a wire or component onto a test position and pull force is measured to remove the wire or component from the test position. This way, the manufacturer can test the strength of the solder without destroying the circuit board. Suppose that the manufacturer properly ran an experiment based upon the following research hypothesis:

> The pull force required to break the soldering test position connector is greater for solder obtained from Acme Wire than for solder obtained from EZ Metals.

Next, suppose that if the experiment had been run correctly, the data obtained *would have* appeared as follows:

EZ Metals	Acme Wire
12.45	16.44
12.32	16.02
12.87	16.32
12.22	16.15
12.76	16.77
12.54	16.57
12.48	16.29

After the hypothetical collection of these data, the statistical analysis would have provided us with the output in Fig. 4.16 and the conclusion that there was a significant difference between the two vendors.

Now, to understand the confounding effect of an extraneous variable in this quadrant of the matrix (see Fig. 4.1), assume that the experiment was run correctly up to the point where measurement of the experimental units took place. Assume that the 14 experimental units were sent to the test laboratory, organized in two boxes of seven boards, each depending on the solder vendor employed. Now, suppose that, because of the inappropriate control of the

SPSS for MS WINDOWS Release 6.1

t-tests for Independent Samples of SUPPLIER

Variable	Number of Cases	Mean	SD	SE of Mean
FORCE Pull Force to Break Solder				
EZ Metals	7	12.5200	.230	.087
Acme Wire	7	16.3657	.253	.096

Mean Difference = -3.8457

Levene's Test for Equality of Variances: F= .082 P= .780

t-test for Equality of Means

Variances	t-value	df	2-Tail Sig	SE of Diff	95% CI for Diff
Equal	-29.76	12	.000	.129	(-4.127, -3.564)
Unequal	-29.76	11.89	.000	.129	(-4.128, -3.564)

Figure 4.16 *t*-Test Comparing the Mean Pull Force Required to Break Solder for Two Solder Suppliers (EZ Metals and Acme Wire).

experimental area, the box corresponding to EZ Metal boards was assigned for testing on test stand 1; the second box (boards produced with solder provided by Acme Wire) was assigned to test stand 2. Unfortunately, test stand 1, when the analysis was performed, was out of calibration by +4 units. In other words, an average of 4 units would have been added to each EZ Metal "true value." This might result in a new set of data to be analyzed:

EZX Metals	Acme Wire
16.45	16.44
16.32	16.02
16.87	16.32
16.22	16.15
16.76	16.77
16.54	16.57
16.48	16.29

Statistically analyzing these data would result in the output in Fig. 4.17. Now the conclusion would be that there is insufficient statistical evidence to infer that there is a significant difference between the two vendors. As shown

Variable	Number of Cases	Mean	SD	SE of Mean

FORCE Pull Force to Break Solder

EZ Metals	7	16.5200	.230	.087
Acme Wire	7	16.3657	.253	.096

Mean Difference = .1543

Levene's Test for Equality of Variances: F= .082 P= .780

t-test for Equality of Means					95%
Variances	t-value	df	2-Tail Sig	SE of Diff	CI for Diff
Equal	1.19	12	.256	.129	(-.127, .436)
Unequal	1.19	11.89	.256	.129	(-.128, .436)

Figure 4.17 t-Test Comparing the Mean Pull Force Required to Break Solder for Two Solder Suppliers (EZ Metals and Acme Wire).

by the comparison of the two analyses, the confounded variable (test stand) has had the effect of masking the true treatment effect (solder vendor). As we have observed throughout this discussion, it is absolutely essential that all experimental designs be carefully constructed and controlled.

Let us return to our potential outcome matrix (see Fig. 4.1), and review the content of the final two cells.

Power: The treatment is related to/does affect the dependent variable, and the experiment results in the identification of the effect

There are two possibilities associated with the condition where the treatment is related to/does affect the dependent variable, and the experiment results in the identification of the effect, called *power* (upper left-hand corner of Fig. 4.1). One, the desired outcome, is that the effect is properly identified, both in terms of its presence and its strength or importance. The second possibility is that the effect is properly identified, but that the strength or importance of the effect is underestimated. This underestimation will generally result from either:

1. The estimation of a treatment effect that is weaker than that which truly exists, due to the testing of levels at a narrower level than appropriate
2. The inflation of an experimental error term, due to the following conditions, singly or in combination:
   a. A weak extraneous variable confounded in the denominator

b. The selection of an incorrect statistical test for the analysis of the data

c. A general lack of sensitivity for the model selected

This last condition, of course, does not result in a false signal, but may result in the misplaced assumption that the treatment effect is less important or influential than it truly is.

Confidence: The treatment is not related to/does not affect the dependent variable, and the experiment results in this conclusion

The condition where the treatment is not related to/does not affect the dependent variable, and the experiment results in this conclusion (called *confidence*), while not necessarily desirable in terms of result we might have hoped to find, is nonetheless a desirable outcome of a well-designed, properly controlled experiment.

As a summary of the principles and techniques we have been discussing, Table 4.3, adapted from Natrella's (1963) work, combines all of these principles in a single table.

TABLE 4.3 Some Requisites and Tools for Sound Experimentation

	Requisites		Tools
1.	The experiment should have carefully defined objectives.	1.	The definition of objectives requires all the specialized subject matter, knowledge of the experimenter, and results in such things as: a. Choice of factors, including their range. b. Choice of experimental materials, procedure, and equipment. c. Knowledge of what the results are applicable to.
2.	As much as possible, effects of factors should not be obscured by other variables.	2.	The use of an appropriate *experimental design* helps to free the comparisons of interest from the effects of uncontrolled variables, and simplifies the analysis of results.
3.	As much as possible, the experiment should be free from bias (conscious or unconscious).	3.	Some variables may be taken into account *by planned grouping*. For variables not so addressed, use randomization. The use of *replication* aids *randomization* to do a better job.
4.	Experiment should provide a measure of precision (experimental error).	4.	*Replication* provides the measure of precision; *randomization* assures validity of the measure of precision.
5.	Precision of the experiment should be sufficient to meet objectives set forth in requisite 1.	5.	Greater precision may be achieved by: refinements of technique, *experimental design* (*including planned grouping*), and *replication*.

Reprinted with permission from Natrella (1963).

Threats to External Validity

All of the threats to validity discussed to this point relate predominantly to internal validity. In the design of any experiment, we must also be concerned with the external validity of the study. External validity refers to the ability to generalize the results of the study to subjects or experimental units outside of the research population, that is, to the target population or universe. Experiments that are internally valid can always be generalized to some extent. Certainly, this ability to generalize extends from the sample drawn to the research population from which the units were drawn. While internal validity is a prerequisite for external validity, it is not necessarily sufficient. There are four general categories corresponding to threats to external validity (Campbell and Stanley, 1963):

1. *Interaction of testing and the treatment (X)*. The interaction of testing and the treatment relates to those designs involving a pretest, where the results obtained for experimental units so treated may not be generalized to the universe of experimental units not pretested.

2. *Interaction of selection and the treatment (X)*. The interaction of selection and treatment refers to the effects of unknown selection biases, and the treatment employed. A major source of this effect in industry is a lack of population validity. This relates to the possibility that a process that is not running in a state of statistical control may not respond to the treatment tested in the same fashion as subsequent populations. Some authors relate this issue to the difference between enumerative and analytical research. Simply stated, research conducted on a process that is not in a state of statistical control may be generalized for the research population studied at the time of the research effort, but if that research population is not representative of some larger universe, the results and conclusions may be totally nongeneralizable.

3. *Reactive arrangements*. Reactive arrangements are of concern in those experiments where subjects are studied. This condition refers to the case where subjects realize that they are part of an experiment, and change their behavior in some way as a result. This change in behavior can affect the dependent variable, and become confounded with the treatment for the purpose of generalization to the larger universe. The so-called Hawthorne effect is one example of a reactive arrangement. A subset of reactive arrangements includes novelty effects, where the observed result is due to the novelty of the treatment for the subjects, rather than the treatment itself. Much of the research in industry conducted during the early days of quality circles suffered from this threat to external validity.

4. *Multiple treatment (X) interference*. Multiple treatment interference refers to the inability to generalize the results of a study from experimental units or subjects which are subjected to multiple treatments, to those universe units which are not equivalently exposed. The multiple treatments might relate to the same effect, or different effects tested through time. This is of par-

ticular concern when subjects are used repeatedly in situations such as a focus group, where multiple testing or treatment exposure in the past may skew the results of the study. The debacle in American education in the attempted application of "classrooms without walls" is another example of this threat to external validity. The majority of the research conducted relative to this application was staged almost exclusively within university and college campuses, among students (subjects) who were as close to professional test subjects as one can get. A treatment that was effective with this population was later found to not generalize to the same degree to the outside universe of public schools.

Unlike threats to internal validity, these conditions cannot necessarily be handled simply by the selection and structure of an experimental design. Rather, in the identification of the target population, the delineation of the statement of the problem, and the construction of the sampling plan, all of these threats must be considered. Appendix B presents a summary table from Campbell and Stanley (1963) for a number of one-factor completely randomized designs, and the threats to internal and external validity against which each design provides protection.

Types of Experimental Designs

We have previously detailed the methods associated with the acquisition of knowledge prior to the advent of the scientific method (the appeals to common sense, authority, intuition, etc.). You will also recall that, as the scientific method became a basis for research, the trial-and-error approach was employed. Following this period, the scientific method was incorporated into the traditional, or one-factor-at-a-time, method of experimental design. The purpose of the content presented here is to describe the structure and nature of the modern method of experimental design. Table 4.4 (Ford Motor Company, 1972) provides a comparison of the two approaches, justifying the use of the more powerful modern methodology.

As we begin to define the designs currently available to us, we note that the classification of designs may be provided in a number of ways:

1. By the number of experimental effects or factors to be investigated (i.e., single-factor versus multiple-factor designs)
2. By the structure of the experimental design (e.g., randomized versus blocked designs)
3. By the nature of the information that the experiment is designed to deliver (e.g., estimates of effects or estimates of variability)

Note that we have already discussed three forms or types of experimental designs: a pretest-posttest control group design, a Solomon four-group design,

TABLE 4.4 A Comparison of Traditional and Modern Methods of Experimental Design

Criteria	Traditional	Modern
Definition of Problem	Objective of experiment roughly understood. The measurements and controls taken largely for granted.	Objective of experimental program fully discussed. The nature of the measurements, and the various levels of the controlled variables, to be studied, discussed and agreed upon.
Experimental Approval	One factor at a time; that is, all variables are held constant except the one being investigated. Once the response is measured, a second variable is then separately studied, holding all other variables constant. Process is repeated for each variable.	Concept of multifactor experimentation used. By running patterns of experiments formed from simultaneous changes in the levels of two (or more) factors great economies in the number of experiments is almost always possible.
Experimental Environment	Care is exercised to control raw material supplies, instruments, men, and all other conditions surrounding the experiments.	In addition to the usual care, the experimenter identifies the variables that contribute to experimental variability and arranges ("blocks") his experimental program to eliminate these contributions. Two types of variables enter most experimental programs: the variables under study, and the blocking variables.
Experimental Error	Presence of experimental error is recognized.	The experimental program is arranged to provide a quantitative measure of the experimental error variance.
Sequence of Experiments	Experiments run in a fixed sequence, or the sequence of trials is casually decided upon.	Randomization of the sequence of experimental trials is formally included as part of the experimental program. Blocks of experiments run in sequence so that early termination of program is always possible.
Number of Experiments	What time and money will permit.	What time and money will permit coupled with an awareness of how large a shift in the response variables is important and the size of the experimental error variance. Experimental designs are employed to enhance the amount of information contributed by each experiment.
Interpretation of Results	Data plots provided, averages computed, and the "obvious" results reported.	In addition to the usual analysis, the variance of the observations is reported, the confidence limit for important effects computed, or in a multivariable experiment, the interaction and second order effects are estimated and least squares estimation procedures used in constructing curves. When appropriate experimental designs are used, great simplifications in computations are possible.
Experimental Success	The success of the experimental program depends on the prior knowledge and ingenuity of the experimenter.	The success of the experimental program depends upon the prior knowledge and ingenuity of the experimenter, augmented by the resources of modern experimental design and analysis.

and a posttest-only control group design. These three designs are simply variations of a single-factor completely randomized design. There are, however, many more types or forms of experimental designs of much greater breadth and complexity. As we begin to utilize these more complex forms, the nomenclature used for the simpler forms is best abandoned in favor of other applicable nomenclature. Even Campbell and Stanley (1963) admit that the attempt to extend the use of their nomenclature to these forms can result in the cre-

TABLE 4.5 A General Outline of Experimental Designs.

Experimental Design	Number of Treatment Variables
I. **Systematic designs**	—
II. **Randomized designs**	—
A. **Complete block designs**	Single
1. Completely randomized design	
2. Randomized block design	
3. Latin Square design	
4. Graeco-Latin Square design	
5. Hyper-Graeco-Latin Square design	
B. **Incomplete block designs**	Single
1. Balanced incomplete block designs	
2. Youden Square balanced incomplete block design	
3. Partially balanced incomplete block design	
C. **Factorial designs**	Multiple
1. Completely randomized factorial design	
2. Randomized block factorial design	
3. Completely randomized hierarchal (nested) design	
4. Completely randomized partial hierarchal (nested) design	
5. Split-plot design	
6. Randomized block completely confounded factorial design	
7. Randomized block partially confounded factorial design	
8. Latin Square completely confounded factorial design	
9. Completely randomized fractional factorial designs	
a. Extreme screening designs	
b. Orthogonal array designs	
c. High resolution fractional factorial designs	
10. Randomized block fractional factorial design	
11. Latin Square fractional factorial design	
12. Graeco-Latin Square fractional factorial design	
D. **Analysis of covariance experimental designs**	Single and Multiple
1. Completely randomized analysis of covariance design	
2. Randomized block analysis of covariance design	
3. Latin Square analysis of covariance design	
4. Completely randomized factorial analysis of covariance design	
5. Split-plot factorial analysis of covariance design	

From *Experimental Design: Procedures for the Behavioral Sciences*, by R.E. Kirk. Copyright© 1995, 1982, 1968 Brooks/Cole Publishing Company, Pacific Grove, CA 93950, a division of International Thomson Publishing Inc. By permission of the publisher.

ation of "traumatizing mysteries" rather than functional descriptions and applications. The many designs reviewed to this point, and the issues surrounding them, should not be considered irrelevant to the more advanced and complex designs. The principles, threats, and concerns previously reviewed with these more basic designs absolutely apply directly to all of the advanced designs. All of the threats to internal and external validity, as well as the utilization of replication and randomization, are inherent considerations in any experimental design, regardless of its complexity.

Kirk (1968) provides one of the more effective breakdowns of the types of experimental designs. We will use this framework for our discussion and description of the forms of modern experimental designs. This breakdown is presented in Table 4.5.

As pointed out by Kirk (1968), systematic designs have been included on this breakdown for historical reference only. The utilization of nonrandom design techniques is no longer of widespread use in industry or agriculture. This is, of course, a result of the inability of the researcher or statistician to apply the techniques of inferential statistics to the results of the study, and, therefore, the fact that the designs lack any significant form of power.

While the major forms of each of the randomized designs will be reviewed in Chap. 11, Table 4.6 provides a brief description of those designs used most frequently in industry today.

TABLE 4.6 Classification and Description of Some Major Experimental Designs

Design	Type of Application	Structure	Information Sought
Completely Randomized	Appropriate when only one experimental factor is being investigated and when material is homogeneous and background conditions can be controlled	**Basic:** One factor is investigated by allocating experimental units at random to treatments (levels of the factor) **Blocking:** None	1. Estimate and compare treatment effects 2. Estimate precision
Randomized Block	Appropriate when one factor is being investigated and experimental or environment can be divided into blocks or homogeneous groups	**Basic:** Each treatment or level of factor is run in each block **Blocking:** With respect to one other experimental variable	1. Estimate and compare effects of treatments free of block effects 2. Estimate block effects 3. Estimate precision
Latin Square	Appropriate when one primary factor is under investigation and results may be affected by two sources of non-homogeneity. It is assumed that no interactions exist	**Basic:** Two cross groupings of the experimental units are made corresponding to the columns and rows of a square. Each treatment occurs once in every row and once in every column. Number of treatments must equal number of rows and number of columns. **Blocking:** With respect to two other variables in a two-way layout	1. Estimate and compare treatment effects, free of effects of the two blocked variables 2. Estimate and compare effects of the two blocked variables 3. Estimate precision
Youden Square	Same as Latin Square but number of rows, columns, and treatments need not be the same	**Basic:** Each treatment occurs once in every row. Number of treatments must equal number of columns **Blocking:** With respect to two other variables in a two-way layout	Same as Latin Square
Balanced Incomplete Block	Appropriate when there is one primary factor but all the treatments cannot be accommodated in a block	**Basic:** Prescribed assignments of treatments to blocks are made. Every treatment will appear at least once in the same block with every other treatment **Blocking:** With respect to one other experimental variable	1. Same as randomized block design but all treatments are not estimated with equal precision
Partially Balanced Incomplete Block	Appropriate if a balanced incomplete block design requires a larger number of blocks than is practical	**Basic:** One primary experimental factor. Prescribed assignments of treatments to blocks are made **Blocking:** With respect to one other experimental variable	1. Same as randomized block design but all treatments are not estimated with equal precision

TABLE 4.6 Classification and Description of Some Major Experimental Designs (*Continued*)

Factorial	Appropriate when several factors are to be investigated at two or more levels and interaction of factors may be important	**Basic:** Several factors are each investigated at several levels by running all combinations of factors and levels **Blocking:** None	1. Estimate and compare effects of several factors 2. Estimate possible interaction effects 3. Estimate precision
Nested	Appropriate when objective is to study relative variability instead of mean effect of sources of variation (e.g., variability of tests on the same sample and variability of different samples)	**Basic:** Factors are strata in some hierarchical structure; units are tested from each stratum	1. Relative variation in various strata
Blocked Factorial	Appropriate when number of runs required for factorial is too large to be carried out under uniform conditions	**Basic:** Full set of combinations of factors and levels is divided into subsets so that some high-order interactions are equated to blocks. Each subset constitutes a block. All subsets are run **Blocking:** Blocks are usually units in space or time. Estimates of certain interactions are sacrificed to provide blocking	1. Same as factorial except certain high-order interactions cannot be estimated
Fractional Factorial	Appropriate when there are many factors and levels and it is impractical to run all combinations	**Basic:** Several factors are investigated at several levels but only a subset of the full factorial is run **Blocking:** Sometimes possible but not necessary	1. Estimate and compare effects of several factors 2. Estimate certain interactions effects (some cannot be estimated) 3. Small fractional factorial designs will not estimate precision

Reprinted with permission from Juran, Gryna, and Bingham (1979).

Chapter 5

Steps for Designing and Assessing an Industrial Experiment

Appropriate Steps for Designing an Industrial Experiment

We have discussed the logic and purpose of an experimental design, the threats to be avoided, and the major forms of experimental designs available to us. This allows us to effectively explore the procedures to be employed in designing valid, reliable, and powerful experiments in the industrial setting. Note that we do not approach the experimental design process by initially deciding what type of design will be selected. Rather, we first identify the independent variables that will be either investigated or appropriately managed, on the basis of the research questions or hypotheses posed. At this point, the type of experimental design appropriate to the research requirements is selected, modified as necessary, and executed.

The following procedure for designing an experiment will be applicable to most situations encountered in industrial settings:

Step 1: Define the statement of the problem.
Step 2: Specifically describe the target population.
Step 3: Define the dependent variables and their associated criterion measures.
Step 4: Identify and classify the independent variables:
　　　　Column 1: Known, manipulable treatment variables
　　　　　　　■ Incorporation
　　　　　　　■ Nesting
　　　　Column 2: Known, manipulable nuisance variables
　　　　　　　■ Limiting/controlling
　　　　　　　■ Blocking

             ▪ Assumption
Column 3: Known, nonmanipulable nuisance variables
             ▪ Random sampling, assignment, ordering
             ▪ Monitoring
Column 4: Unknown, extraneous effects
             ▪ Random sampling, assignment, ordering
Step 5: Select the levels associated with each treatment variable.
Step 6: Select the experimental design appropriate to the treatment variables and number of levels selected. Modify as required and assign the treatments to the design plan.

Steps 1 and 2 were discussed in previous chapters. In this chapter, we will explore Steps 3 through 6.

Step 3: Define the dependent variables and their associated criterion measures

Generally, the dependent variables associated with a given experiment are well defined and understood. (There may, of course, be only one dependent variable and one criterion measure.) This will relate to the background database, and the reason that the study was instigated in the first place (refer to Chap. 1 for a discussion of this premise). A major error that can occur at this juncture, however, relates to the tendency of novice practitioners to exclusively focus on the dependent variables of interest, and their associated criterion measures.

For example, imagine that a problem-solving team (PST) is working on the problem of ingot cracking which is the leading cause of internal scrap (cost). The team has determined that the current cracking rate, running at an average of 12 percent per month, is in a state of statistical control. This means they have determined that they must pursue the common causes of variability in the system. They determine that the certification of process design quality is the requisite step in their effort to eliminate this internal dissatisfier. Forming an experimental design team for this purpose, the group designs and executes an experiment to determine the root causes associated with ingot cracking.

Five of the factors studied, including two interaction effects, turn out to account for the majority of the influence associated with ingot cracking. Verification experiments show that the initial study possesses external validity. The PST modifies the standard operating procedures (SOPs) associated with the casting process in order to institutionalize the findings of the research. The ingot cracking rate falls to 0.00 percent, as long as process control is maintained around the established targets for the first- and second-order critical process variables. Everyone is delighted.

Four weeks later, the plant manager receives what can only be described as a catastrophic telephone call from the major customer of the plant's output. As a result of a great deal of subsequent analysis, it becomes apparent that the changes to the casting process had a deleterious effect on critical in-process product characteristics generated five production centers downstream from the casting process. The changes in these critical characteristics in the post-experimental production ingots were not initially detected, because they were not measured on an ongoing basis. They were not measured as a normal part of the production sequence, because they had long ago been brought into a state of statistical control at capability indices in excess of 6.00.

The point of this case study is to underscore that no experiment should be conducted in industry that is limited, intentionally or by default, to only the criterion measure of immediate interest. This approach will potentially yield one of three outcomes:

1. The suboptimization of the process or product
2. A catastrophic effect on downstream internal and/or external customers
3. A lost opportunity for process improvement resulting from research conducted on a dependent variable or criterion measure that was not the focus of the original study

In other words, it is entirely possible that if we conduct a study on one dependent variable, and then follow the product through subsequent processes, we might find that the treatment effects manipulated for the experiment have a positive effect on other downstream variables. If the criterion measure of original interest is the only variable measured, and if the treatment shows no effect on this measure, the opportunity for improvement could be lost!

In summary, therefore, all dependent variables, and their associated criterion measures, that might conceivably be affected by the treatment variables should be identified by the design team and measured as part of the experiment. This should be done in every case, unless significant economic considerations are involved. In these cases, it should be noted that subsequent verification experiments will be mandatory to assess the potential interactive effects of the modified process on the critical variables not measured during the experiment. It must be noted, however, that this procedure will guard only against catastrophic events. It will not provide us with measures of potential benefit in those cases where the treatment variables do not affect the criterion measures of primary interest.

To assist in the development of the experimental design at this juncture, Fig. 5.1 is provided. This figure will be incorporated into our developing checklist.

Step 4: Identify and classify the independent variables

If there is one critical element of sound experimental design, it is that independent variables must be properly managed if we are to execute an accurate and precise research study. This procedure requires that we understand the implications (in terms of the cost and complexity of the resultant design) of the

108   Chapter Five

Primary Dependent Variable(s)	Criterion Measure(s)	Measurement Scale
1.	1. 2. 3. 4. 5.	
2.	1. 2. 3. 4. 5.	
3.	1. 2. 3. 4. 5.	

Secondary Dependent Variable(s)	Criterion Measure(s)	Measurement Scale
1.	1. 2. 3.	
2.	1. 2. 3.	
3.	1. 2. 3.	

Figure 5.1 A Preliminary Breakdown of Dependent Variables and Criterion Measures for Inclusion in the Experimental Design.

options available to us for the manipulation of prospective independent variables. As the first activity in this step, the design team lists all of those independent variables which are known to, or are suspected to, affect the primary dependent variables. Independent variables suspected to affect only the secondary dependent variables should not be listed, at least initially. Next, the independent variables should be organized according to whether they are manipulable or nonmanipulable. Two observations are made at this point:

1. The design team should (generally) include a cross-functional representation of those individuals knowledgeable about and familiar with the product or process to be studied. Production managers and supervisors, engineering, maintenance, and line personnel (hourly employees) should be included on the team so that potential confounding effects will not be overlooked.

2. Begin the design process using rules similar to brainstorming. That is, do not fail to record any independent variable suggested. Remember, randomizing across significant extraneous variables undermines the sensitivity of the final design. Even if we do not include the variable in the study, it will still have to be monitored in some way.

We now have two columns of identified independent variables. Turning to the column associated with the manipulable independent variables, we must now decide which of these factors will and will not be included in the research study. This is not a small consideration. As each independent variable is added to the study, transforming the factor from a nuisance variable to a treatment variable, the experiment becomes larger, more complex, and more expensive. On the other hand, eliminating independent variables from inclusion in the experiment may threaten the internal or external validity of the research. In this context, the design of an experiment requires us to establish a compromising balance between the knowledge that we would like to acquire and the cost of the acquisition.

Based upon the statement of the problem, and the research questions or hypotheses that have been developed, there are some independent variables that would naturally be included for study in the design at this point. The consideration of the additional independent variables identified by the design team may require a form of catchball between the team, the sponsor of the experiment, and the management team. Basically, the question to be answered is

Is this a variable that we want information about?

If the answer is yes, then the variable must be included in the study, at the expense of size, complexity, and time. As these additional factors are added to the experiment as treatment variables,* the list of research questions or hypotheses should be augmented. There should be no need to modify the statement of the problem since we are simply identifying and studying additional sources of variability that are of interest, but secondary to the original purpose for conducting the experiment.

Independent variables that are manipulable, but will not be included in the experiment, should be organized in a second column. These factors must still be properly managed in the conduct of the experiment, however. Nuisance variables, known but nonmanipulable independent variables, may not be (strictly speaking) managed in the conduct of the study, but must not be ignored. Finally, independent variables that are unknown to the experimental design team constitute a second set of nuisance variables.

---

*Some authors might refer to the primary independent variables of interest as the "true" treatment variables. Additional factors added for study would simply be referred to as *independent variables* or *covariates*; all other factors or effects would be referred to as *nuisance* variables.

TABLE 5.1 Options for the Management of Independent Variables in an Experimental Design.

Independent Variable Classification	Known			Unknown
	Manipulable		Nonmanipulable	
	Study	Don't Study		
Method	Incorporate / Nest	Limit/Control / Block / Use Assumption(s)	Randomization with Monitoring	Randomization
CONSEQUENCES Sample Size (n)	Increase	No Increase	No Increase	No Increase
Design Complexity	Increase	Increase for Blocking / No Change for Other Alternatives	No Change	No Change
Confounded Denominator	No	No for Blocking and Control / Possibly Yes for Assumptions	Probably	Probably

In order to discuss the various options available for each of these categories, Table 5.1 is provided. This table will be utilized to coordinate our discussion of the management options and consequences associated with the treatment of the independent variables in an experimental design.

Column 1: Known, manipulable treatment variables. Once we have decided that we will study a particular independent variable, we have two options for the method by which we will arrange or organize the variable in our experiment: incorporating or nesting.

Incorporating independent variables. When we speak of incorporating a treatment variable, we are implying that we will study and report on that factor across all other variables and effects in the experiment. Suppose, for example, that a company manufactures a product with a critical characteristic of strength as measured by tensile yield strength and the company wishes to test the following research hypotheses:

1. There is no significant difference between the average end-of-line tensile yield strength (TYS) for product manufactured with material supplied by our three current vendors (Acme, Southco, and Newmark).
2. There is no significant difference between the average end-of-line TYS for product manufactured on production line 1 and production line 2.

Constructing a single experiment to simultaneously test both of these treatments (vendor and production line), we observe that each of the hypothesis statements relates to the expected outcome of a factor comparison across a second factor. In other words, when we test the hypothesis related to vendor dif-

		Treatment Variable 'A' Vendor		
		Acme	Southco	Newmark
Treatment Variable 'B'	#1			
Production Line	#2			

Statistical hypotheses to be tested:

$H_0: \mu_{ACME} = \mu_{SOUTHCO} = \mu_{NEWMARK}$
$H_1: \mu_{ACME} \neq \mu_{SOUTHCO} \neq \mu_{NEWMARK}$

$H_0: \mu_{LINE1} = \mu_{LINE2}$
$H_1: \mu_{LINE1} \neq \mu_{LINE2}$

$H_0: I_{AB} = 0$
$H_1: I_{AB} \neq 0$

Where each cell contains n sample observations.

Figure 5.2 Sample Experimental Design Layout for Two Treatment Variables (Nested Design).

ferences, we are testing whether the average end-of-line TYS varies by vendor across, or regardless of, production lines. When we test the line effect, we are implying that we wish to compare the two lines across, or including, material supplied by all three vendors. As we will review in Chap. 11, this type of requirement will lead us to design a 2×3 factorial experiment. Additionally, we will have to test one additional hypothesis to conduct this study. This additional hypothesis will relate to the assessment of an interaction effect. This effect and the associated testing procedure will be reviewed in Chap. 11. For now, you should observe that when we have two treatments that are both incorporated, we have created a *crossed* analytical model. This term relates to the fact that the treatments are intended to be analyzed *across* all other treatments in the design. This design intention is displayed in Fig. 5.2.

The statistical hypotheses to be tested associated with the model are

$H_0: \mu_{ACME} = \mu_{SOUTHCO} = \mu_{NEWMARK}$

$H_1: \mu_{ACME} \neq \mu_{SOUTHCO} \neq \mu_{NEWMARK}$

$H_0: \mu_{LINE1} = \mu_{LINE2}$

$H_1: \mu_{LINE1} \neq \mu_{LINE2}$

$$H_0: \quad I_{AB} = 0$$
$$H_1: \quad I_{AB} \neq 0$$

where each cell contains n sample observations.

Nesting independent variables. When we speak of including an independent variable in an experiment as a treatment variable in a nested context, we are essentially referring to the comparison of levels or groups *within* levels of a second factor, as opposed to across levels of a second factor. This type of design, often referred to as *hierarchical* or *hierarchial*, is employed when one of two conditions exist:

1. The nature of the research question requires that the comparison of the treatment effects takes place within versus across the levels of a second factor.
2. The assessment of the treatment effect makes no logical sense when analyzed across levels of a second treatment variable.

For example, a steel rolling mill operates a slitter at the end of one of its production lines. These slitters use a series of finely ground knives to slit, or cut, the final product. One important point related to these purchased knives is that they are very expensive to purchase and replace. The steel company's slitting department manager wants knives that not only work well on her slitter, but that will last a long time (life) before requiring a production shutdown to sharpen or change the knives.

Currently, assume that the steel company purchases these knives from a supplier we will refer to as Abbot Knife Works. Imagine that two additional suppliers have asked us to review and qualify their knives for possible purchase in the future. As an additional aspect of this issue, assume that our current supplier, Abbott, and the two potential suppliers, Ranco and Charlesco, each have five production plants which produce these knives. If we purchase from any single supplier, it is possible that the knives we receive may come from any of the plants associated with the selected supplier. What concerns us here is that we want to compare not only the suppliers, but the variability of knife quality and life from plant to plant, in case each plant does not produce the same quality of product. Ideally, the best purchasing decision would be to purchase the knives from the supplier that offers the best combination of cost and quality, where there is no difference in knife quality or life, no matter which of their production facilities produced the knives delivered to the slitting department.

In this case, the treatment variable, supplier, would be incorporated for study. This decision implies that the supplier effect will be assessed across all levels of plant included in the study.

Consider, however, the second treatment effect, plant. If we have decided that we wish to test for the variability of the five plants associated with each

Designing and Assessing an Industrial Experiment 113

```
                        Treatment Variable 'A'
                              Vendor

                    Abbott      Ranco      Charlesco
              #1  ┌─────────┬──────────┬───────────┐
                  │         │          │           │
Treatment     #2  ├─────────┼──────────┼───────────┤
Variable 'B'      │         │          │           │
              #3  ├─────────┼──────────┼───────────┤
  Plant           │         │          │           │
              #4  ├─────────┼──────────┼───────────┤
                  │         │          │           │
              #5  ├─────────┼──────────┼───────────┤
                  │         │          │           │
                  └─────────┴──────────┴───────────┘
```

Where each cell contains n sample observations.

Figure 5.3 Sample Experimental Design Layout for Two Treatment Variables (Nested Design).

supplier in the study, we must also note that the levels corresponding to this variable are not logically comparable. That is Plant 1 is simply a nominal assignment. The categorization of any particular plant as number 1 is an arbitrary numerical assignment. In a crossed context, therefore, assessing Plant 1 versus Plant 2 versus Plant 3 versus Plant 4 versus Plant 5 is meaningless. The experimental design, therefore, is modified from a crossed model to a nested model (for studies with more than two factors, designs are referred to as fully or partially nested). Figure 5.3 provides an example of how this design might appear given the constraints discussed to this point. Additional applications of this principle will be presented in Chap. 11.

The following statistical hypotheses would need to be tested:

H_0: $\mu_{ABBOTT} = \mu_{RANCO} = \mu_{CHARLESCO}$

H_1: $\mu_{ABBOTT} \neq \mu_{RANCO} \neq \mu_{CHARLESCO}$

Within Abbott:

H_0: $\mu_{PLANT1} = \mu_{PLANT2} = \mu_{PLANT3} = \mu_{PLANT4} = \mu_{PLANT5}$

H_1: $\mu_{PLANT1} \neq \mu_{PLANT2} \neq \mu_{PLANT3} \neq \mu_{PLANT4} \neq \mu_{PLANT5}$

Within Ranco:

H_0: $\mu_{PLANT1} = \mu_{PLANT2} = \mu_{PLANT3} = \mu_{PLANT4} = \mu_{PLANT5}$

H_1: $\mu_{PLANT1} \neq \mu_{PLANT2} \neq \mu_{PLANT3} \neq \mu_{PLANT4} \neq \mu_{PLANT5}$

Within Charlesco:

H_0: $\mu_{PLANT1} = \mu_{PLANT2} = \mu_{PLANT3} = \mu_{PLANT4} = \mu_{PLANT5}$

H_1: $\mu_{PLANT1} \neq \mu_{PLANT2} \neq \mu_{PLANT3} \neq \mu_{PLANT4} \neq \mu_{PLANT5}$

Column 2: Known, manipulable nuisance variables. In Table 5.1, Column 2 relates to those known and manipulable independent variables:

- That we do not wish to incorporate or nest in our experiment because they are not integral to our statement of the problem and research hypotheses
- Where the additional cost, time, and expense of gathering data associated with these variables is considered too high given the noncritical nature of the effects
- Where we cannot ignore the effects of these factors and allow them to confound with our treatment effects
- Where we cannot allow them to confound with and artificially inflate our experimental error term, perhaps resulting in a catastrophic reduction in the sensitivity of the design

When we identify these independent variables, we have three choices as to how we will protect our experiment against their potentially confounding and devastating effects: limitation/control, blocking, or assumptions.

Limitation/control. *Limitation,* also referred to by some authors as *control,* is the process by which an extraneous variable is prevented from confounding an experiment by limiting the factor to a single level. For example, suppose that we wished to compare three vendors (our incorporated treatment variable). Suppose further that each vendor might ship us material from one of three independent sources, or plants. Assume, however, that the experiment is expensive to run, and we simply are unwilling to test nine conditions or treatment combinations (the consideration of whether the model will be crossed or nested is, in this case, irrelevant). Management's position is that we simply cannot afford to run that experiment. One method of dealing with this situation would be to select one of the plants from each vendor, and gather all of the data from that source. This process allows us to eliminate the threat that plant-to-plant variability will confound in our experimental error term, in that only one level is tested. This option is deceptively attractive. There have been a number of single-factor experiments conducted by uneducated personnel where virtually all extraneous manipulable variables have been limited. Of course, this could lead to an experiment on vendor material for (this will be an extreme example) line 6, in plant 3, on the day shift, for bodymaker 3, on Mondays. While this type of study would certainly protect us from a number of extraneous variable effects, there is no free lunch. The caveat against the indiscriminate use of limitation as a management option for a nuisance variable is that we may finish with conclusions that have little or no external validity, due to a concern for population validity. That is, given these conditions, you may not, under any circumstances, make inferences to any variable level other than the level studied within the experiment.

If, for example, we restrict our vendor experiment to a single plant, and if we subsequently conclude that vendor differences exist, we may only generalize those results to the individual plant tested. When might this be appropriate? Suppose that for each vendor, we were receiving 95 percent of the input from only one of their respective sources. In this case, the inability to make an inference to the other sources, given the additional expense, might be considered inconsequential. This, of course, assumes that the incoming mix is not expected to change, or is expected to become even more heavily sourced in the plant to which the experiment is limited.

If this is not the case, limitation is not considered a sensible option for the management of the extraneous variable.

Blocking. If, because of the shortcomings associated with the option of limitation, we cannot limit the design to a single level of an extraneous independent variable, then we will typically block out its effects. By *blocking,* we are referring to the practice of constructing a design such that:

1. All levels of the variable necessary for external validity are included in the design
2. The inclusion does not require additional replication
3. The blocked variable cannot become confounded with the treatment effect
4. The variability associated with the blocked variable is subtracted, or not permitted to confound with, the experimental error term

Blocked designs will be explored in depth in Chap. 11. There is one caveat associated with blocking that should be noted at this point, however, and it will be repeated later. When we employ blocking as a management option, we are assuming that there is no interaction between the treatments and the blocked variable; that is, the variability associated with the blocked factor is uniform across all levels of the treatment variable. If this is not the case, then the nature of the research question must be such that an interactive effect is unimportant to the research hypothesis; i.e., that a compromise or "across-block" answer is desired.

If there is an interaction, and if the research question does not support an across-block analysis, then the variable should not be blocked in the first place. Often, statisticians will get into in-depth and emotional discussions on this topic as associated with the assignment or decomposition of variability terms. From a researcher's point of view, they have missed the point. A blocked variable is not a treatment variable. If this issue is even being discussed, the original design approach was significantly flawed.

Assumption. While not a recommended approach, its widespread use requires that we mention assumption as an option for the management of extraneous independent variables. In essence, this option refers to the decision to simply "assume" that independent variables suggested as possible threats to the design simply do not have any relationship to the dependent variables or criterion measures associated with the study. Given that these variables have no effect, the assumption is that there is no sense in worrying about limiting or blocking out their effects. In the opinion of these authors, this method of managing extraneous variables is a recipe for disaster. After all, if everyone correctly knew all of the factors affecting the primary dependent variable, then it is unlikely that we would be conducting the experiment in the first place. As a result, it is recommended that this option never be employed.

Column 3: Known, nonmanipulable nuisance variables. In most industrial experiments, there will be a number of independent variables identified by the design team that are known or suspected to affect the primary dependent variables, but cannot be manipulated by the researchers. (See column 3, Table 5.1.) The biggest mistake novice practitioners make in conjunction with these variables is to assume they must be managed as if they were unknown factors, through the randomization of selection, assignment, and order. While randomization will, in fact, be employed in conjunction with these variables, this is not the only tool to be utilized for these factors. These variables must be appropriately monitored, or tracked, through the conduct of the experiment. The use of a properly constructed experimental or engineering design log will allow us to accomplish this task. An example of this type of log is presented in Chap. 8 and in the appendices. The appropriate monitoring of these variables may be the only method available to the design team to provide a data-based theory associated with the supposed relationship between extraneous, uncontrollable factors and the magnitude of the experimental error term generated in an experiment.

Topic for discussion: How frequently should these variables be measured during an experiment?

Column 4: Unknown extraneous effects. Obviously, if an effect is unknown, it cannot be addressed by the design of the experiment. Random sampling, assignment, and ordering will minimize the effect of any unknown factors.

Classifying the treatment variables. After all these decisions have been made, we move to the process of classifying the treatment variables selected for incorporating or nesting within the experiment. This classification is made on the basis of two attributes: quantitative or qualitative, and fixed or random.

Our first task in the classification of any treatment variable studied in an experiment is to identify whether the underlying distribution associated with a factor or effect is, by nature, qualitative or quantitative. This is essential if we are to properly perform associated statistical analyses at a later point in the research process. A qualitative factor is one where the levels of the variable are finite, categorical conditions. Examples of qualitative factors include effects such as vendor, production line, manufacturing plant, state, shipload, and packaging method. Quantitative variables are those where different treatment levels constitute different amounts or degrees of the factor (Kirk, 1968) and where the measured variable can take on a relatively infinite series of values. Examples of quantitative factors include temperature, pressure, viscosity, speed, feed, and height. Incidentally, do not confuse the classification of a treatment variable as qualitative or quantitative on the basis of the possible limitations of the measurement process. Whether we are supplied with a micrometer or a go–no go gauge, thickness is a quantitative variable, in terms of its underlying distribution.

The second treatment variable classification required is the identification of the factor as a fixed or random effect. A fixed effect is a treatment variable where all levels of the factor of interest to the researcher/experimental design team/management team are included in the experiment. For example, recall the experiment associated with a crossed model, as associated with a vendor analysis. You will remember that there were three vendors (levels) tested in this design. Conceivably, there could be hundreds of vendors who could supply this incoming material. Designation of the factor as a fixed effect implies that the researcher wishes to make inferences to only the levels purposively selected for inclusion in the experiment. Effects associated with other levels (vendors, in this case) are of no interest or consequence, insofar as the experiment is concerned.

A random effect, on the other hand, is one where the researcher has randomly selected the levels of the treatment variable associated with the study. The intention in this case is to provide an inference not to the levels included in the research study, but to generate inferences related to the population of treatment levels represented by the experiment. For example, suppose we were conducting an experiment designed to compare three vendors as an incorporated fixed effect (treatment A) and we also wanted to assess the variability among coils of incoming product supplied by each potential vendor as a nested random effect (treatment B). Figure 5.4 reflects the general structure that this design might take. Take careful note of the change in the statistical hypotheses to be tested.

Statistical hypotheses to be tested:

$$H_0: \mu_{ABBOTT} = \mu_{RANCO} = \mu_{CHARLESCO}$$
$$H_1: \mu_{ABBOTT} \neq \mu_{RANCO} \neq \mu_{CHARLESCO}$$

Within Abbott:

$$H_0: \sigma_T^2 = 0$$
$$H_1: \sigma_T^2 \neq 0$$

		Treatment Variable 'A' Vendor		
		Abbott	Ranco	Charlesco
	#1			
Treatment Variable 'B'	#2			
Coil	#3			
	#4			
	#5			

Where each cell contains n sample observations.

Figure 5.4 Sample Experimental Design Layout for Two Treatment Variables (Nested Design, Model III).

Within Ranco:

$$H_0: \sigma_T^2 = 0$$

$$H_1: \sigma_T^2 \neq 0$$

Within Charlesco:

$$H_0: \sigma_T^2 = 0$$

$$H_1: \sigma_T^2 \neq 0$$

where σ_T^2 corresponds to the treatment effects, the variability associated with the population of treatment variable levels.

Studies with more than a single factor or treatment are categorized (for analytical purposes) as Model I (all fixed effects), Model II (all random effects), and Model III (a mixed group of fixed and random effects) designs.

Step 5: Select the levels associated with each treatment variable

The number and value of levels for each treatment should be selected with as much care and consideration as the treatment variables. Qualitative factors of a fixed nature should include all levels of the factor for which we desire to make an inference. For qualitative and quantitative factors that are, according to the research hypotheses, appropriately treated as random effects, the number of levels randomly selected and tested should be adequate to answer the associated research questions. A statistician can assist with the calculations necessary to determine the minimum number of levels required for the study of quantitative factors. For qualitative factors, a percentage of the available levels in the research population of all levels is often employed.

For quantitative factors treated as fixed effects, the number of levels that should be studied is often based upon whether a nonlinear (quadratic or cubic) effect is expected. Three levels are run to assess for a quadratic effect. Four levels must be run to test for a cubic relationship.

In other cases, where nonlinearity is only a possibility, but where the effect may pass through the target or nominal value for the criterion measures, more than two levels will often be selected. Of course, these considerations have a direct bearing on the cost of the experiment. More levels result in more observations and, therefore, greater complexity of controlling the experiment.

For quantitative treatment variables studied as fixed effects, the extremes (i.e., the high and the low values) of the levels selected should be as far apart as possible without exceeding the operating envelope of the system studied. If additional levels are to be selected outside of this envelope, make certain that (1) all interested parties are apprised of the decision, and agree with it, and (2) the levels selected possess a degree of external validity with the target population.

On the other hand, it is the responsibility of the researcher to verify that the set points or targets selected for the levels to be studied are far enough apart so that the common variability associated with each level is nonoverlapping. In other words, suppose we conduct an experiment where we run the temperature of a plating tank at 135° and 140°. Suppose further that, regardless of central tendency, the variability about the set point may be represented by a normal distribution and has a standard deviation of 3°. This would imply that, approximately 16 percent of the time that the tank is set to 135°, it is actually running at 138° or more. Additionally, when the tank is set to run at 140°, 16 percent of the time it is actually running at 137°, or lower than the supposed "low" set point when it happens to be running high. This condition is unacceptable, in that it would seriously threaten our ability to test for a true temperature effect. Even if we were to increase the replication level to extremely high numbers to offset this condition, our estimate of experimental error is still compromised.

Topics for discussion: What might we do in this case? What are the consequences of these alternatives?

Figure 5.5 provides a sample form which may be employed to describe, organize, and classify the treatment variables incorporated and nested within the experimental design. A section of this form has also been set aside to identify those variables to be blocked and limited.

Finally, never test a factor at a level that, regardless of the result of the research study, will not be employed by the management team. For example, an experiment was conducted, at a significant cost, in conjunction with a manufacturing process where two of the factors studied were speed and cure time. The result of this (technically) excellent study was that the optimum condi-

Treatment Variable(s)	Method I - Incorporated N - Nested	Classification QN - Quantitative QL - Qualitative	Type F - Fixed R - Random	Number of Levels	Level Description
1.					
2.					
3.					
4.					
5.					
6.					
7.					
8.					
9.					
10.					
Blocked Variables	**Number and Value for Levels Blocked**		**Variables Limited**		**At Level**
1.			1.		
2.			2.		
3.			3.		
4.			4.		

Figure 5.5 Sample Table for the Organization of Treatment, Block, and Limited Variables.

tions identified corresponded to, among other settings, the levels of low speed and long cure time, as studied in the experiment. Unfortunately, management had no intention of running the process at that speed, or for that cure time. Either the economic constraints were not well understood by the experimental design team at the outset, or proper communication related to the design of the study was not executed. Therefore, a significant amount of money was spent in this case to understand conditions that had absolutely no external validity. To avoid this possibility, the authors recommend that some form of memorandum be sent to "interested parties" prior to the conduct of an experiment to alert the design team to the potential for this type of condition. Figure 5.6 provides a sample form that may be employed for this purpose.

Step 6: Select the experimental design appropriate to the treatment variables and number of levels selected; modify as required, and assign the treatments to the design plan

In Chap. 11, we will review a series of case studies associated with the major experimental designs often employed in industry. Each experimental design, such as a randomized block design, may appear in different forms, e.g., Latin Squares and Youden Squares. In addition, Latin Squares are available in different sizes. As another example, it may be decided that a fractional factorial design is required; but it may be appropriate to use a high-resolution "classical" design, an orthogonal array (a type of design), or a Plackett-Burman design.

In summary, the key is that there is only one appropriate method for designing an experiment. That is, identify the treatment factors which must be studied to test the research hypotheses, and then design the most efficient, least expensive experiment possible to allow us to test these hypotheses. This analysis must also take place with the power and precision required given the

---

Date: _____

To: _____

From: _____

Subject: Planned Experiment Input and Reaction Request

==

The team working on *(topic and purpose)* has developed a proposed experimental design that is intended to respond to the following Statement of the Problem:

(insert Statement here)

The Research Hypotheses the design is intended to test appear on Attachment A. The Treatment variables we have decided to include in the experiment, and the settings which will be tested are presented in Attachment B.

(In the case of Fractional Factorial Designs, include the following paragraph)

In addition to the main effects to be tested, we are also planning to assess the process for the interaction effects listed on Attachment C.

The experiment is planned for initiation on *(date)*. We expect that it will take approximately *(time)* days to run, and will require assistance from the following personnel: *(specify)*

Please review the scope and structure of the design. If there are any additional factors which should be included, or if you have any concerns as to the settings selected, or the design selected for the conduct of the experiment, please contact *(specify the name of the design team leader)* or *(specify the name of the sponsor)* no later than *(specify final date for input and suggestions)*.

Thank you for your assistance in our Quality Improvement effort.

Figure 5.6 Experimental Design Notification Memorandum.

consequences associated with correct and incorrect conclusions as related to the design. Surprisingly, there are actually some individuals who have learned a few pet designs, and obtained software to process their associated data sets. These individuals regularly embark on searches for research questions that may be answered by these designs, exactly the opposite of the approach that should be utilized.

Assessing the Industrial Experiment for Adequacy and Efficiency

Many authors have suggested questions for determining whether the experimental design selected is appropriate—that is, whether the design is adequate and efficient. These suggested questions may be distilled into five items (Kirk, 1968):

1. Does the design permit the researcher to calculate a valid estimate of treatment and experimental error effects?
2. Does the experiment require and include a data collection procedure that will generate reliable and valid results?
3. Does the design provide a sufficient level of power to permit an adequate test of the statistical hypotheses, at the effect size and Type I and II error levels required?
4. Does the experimental design conform with accepted practices and procedures as associated with the modern approach to experimental design employing the scientific method?
5. Does the design provide maximum efficiency within the required constraints of the experimental area?

The last item, related to an assessment of maximum efficiency, requires additional discussion. When we speak to the *efficiency* of an experimental design, we are referring to the time and cost associated with conducting an experiment, as compared with the breadth and depth of the information acquired. In an effort to describe this comparison in terms of a ratio of information to cost, Cochran and Cox (Kirk, 1968) proposed an efficiency index (EI) for the comparison of two experimental designs. This formula is particularly instructive in light of the components considered. This index appears as

$$EI = \frac{\left(\frac{n_2 C_2}{\hat{\sigma}_1^2}\right)\left(\frac{df_1 + 1}{df_1 + 3}\right)}{\left(\frac{n_1 C_1}{\hat{\sigma}_2^2}\right)\left(\frac{df_2 + 1}{df_2 + 3}\right)}$$

where $\hat{\sigma}^2$ = estimate of experimental error per observation
n = number of experimental units or subjects
C = cost of collecting data per experimental unit or subject
df = degrees of freedom for the experimental error term in the respective design

If the ratio is less than 1, the second alternative design is more efficient than the first. If the ratio is greater than 1, the first design is the more efficient. As

Kirk (1968) points out, however, one cannot simply select a design solely on the basis of this index. One design may be advantageous on the basis of one attribute (say, cost) while an alternative design may be more advantageous on the basis of a second attribute, such as reduced experimental error. The goal before the experimental design team or researcher is to select that valid design which is the most powerful and sensitive available, within the constraints presented.

Self-Review Activity 5.1 You now have an opportunity to organize an experiment. We have noted that an understanding of the research problem, technology, and experimental area are all required to properly design a study.

You are currently a member of a quality improvement team. This team has been working to bring the critical characteristic, sheet cleanliness, into a state of control and capability. This characteristic is an in-process critical product characteristic, measured continuously as the sheet exits the cleaning tank. The measurement process used for this attribute is off line (in the laboratory). There is only one test stand available, but two laboratory technicians perform the tests: Bob and Kathy.

The sheet material in question enters the cleaning tank directly from the incoming dock. There are three vendors for our incoming material: Ajax, Quality, and Foamco. There are two possible threading procedures that could be used for the cleaning process. The critical control parameters (first-order critical process variables) and their specifications are

Temperature: 145 to 165°F

Concentration: 35 to 45 percent

Line speed: 400 to 450 feet per minute

Squeegee pressure: 10 to 15 pounds per square inch

These variables and their settings were supplied by the cleaning process equipment vendor. We have no evidence that the equipment vendor has ever tested these assumptions and no targets or optimal levels have ever been provided to us.

The cleaning process is run on two shifts, the day shift and the night shift. Someday, if we can get our sheet clean, we may be able to justify a graveyard shift, although none currently exists.

Currently, the cleaning process requires the following product output from the cleaning tank:

$$150 \pm 20 \text{ units}$$

Unfortunately, our process is running in a state of control at a mean of 140 and a standard deviation of 10 units. The distribution has been determined to be normal, and to date we have been unable to successfully shift the mean of the process without affecting the variability of the process.

Complete the following tasks:

1. Write a statement of the problem.
2. Write an appropriate set of research hypotheses.
3. Consider all of the independent variables described for inclusion in the experiment.
4. Categorize and classify the treatment variables you select for inclusion in this study (you may use the form in Fig. 5.5 for this purpose). Identify the blocked variables and limited variables associated with your recommended design. Do not attempt to design the experiment itself at this point. You will have an opportunity to design and analyze a similar experiment in Chap. 11.

Updating the Planning Checklist for Industrial Research

The content reviewed in this chapter has been incorporated into the checklist presented at the end of Chap. 1. Review the expanded checklist (Fig. 5.7) and compare the items to the material you prepared for your self-review activity. Have you considered all of the identified and important elements of the design process in preparing your own design plan?

I. Developing the Statement of the Problem

☐ A. A comprehensive and complete background data base for the research study exists, originating from a: (✓ one)

  ☐ Strategic product-market analysis
  ☐ Technical competitive benchmarking study
  ☐ Quality improvement effort
  ☐ Problem-solving effort
  ☐ Other (specify)_____

☐ B. The background data base includes a statement of significance for the problem

☐ C. The originating problem has been properly delineated and delimited

☐ D. A specific and focused statement of the problem has been developed that is: (all should have a ✓)

  ☐ Concise
  ☐ Precise
  ☐ Testable
  ☐ Obtainable

II. Classifying the Research Study

The research design to be conducted has been identified as: (✓ one)

☐ Agreement research
  ○ Consensus
  ○ Instrument/system concordance

☐ Relational research
  ○ Concurrent correlational
  ○ Predictive correlational
  ○ Causal comparative

☐ Descriptive research
  ○ Status study
  ○ Longitudinal study
  ○ Case study
  ○ Cross-sectional study

☐ Experimental research

III. Operationalizing the Statement of the Problem

☐ If the research design is nonexperimental, research questions have been developed; or if the research design is experimental, research hypotheses have been developed

☐ The research questions or hypotheses have been evaluated and assessed as: (all should have a ✓)
  ☐ Clear
  ☐ Testable
  ☐ Specific
  ☐ Simply stated
  ☐ Consistent with known facts

Figure 5.7 A Planning Checklist for Industrial Research.

IV. **Designing the Industrial Experiment** (all should have a ✓)

- ❏ The target and research populations have been described in clear and specific terms

- ❏ The primary dependent variable(s) and the associated criterion measure(s) has been identified, and is consistent with the research hypothesis(es) to be tested. Secondary dependent variables have also been identified. (The first two columns of checklist attachment A have been completed)

- ❏ All independent variables that may have an affect on the primary or secondary dependent variables have been identified

- ❏ The treatment variables have been identified and classified to be incorporated or nested

- ❏ The known and manipulable nuisance variables have been identified and classified to be blocked or limited

- ❏ The known and nonmanipulable nuisance variables have been identified and an appropriate monitoring plan has been set up for each factor for use during the conduct of the experiment

- ❏ The treatment variables have been correctly categorized as:
  * Qualitative versus quantitative
  * Fixed versus random effects

- ❏ Classification of fixed versus random effects for each treatment variable has been shown to be consistent with the statement of the problem and the research hypothesis(es)

- ❏ For each treatment variable, an appropriate number of levels has been identified. Level selections are appropriately within the operating envelope of the system, set at an extreme level, and have been shown to be nonoverlapping

- ❏ A summary table for independent variable management has been completed (checklist attachment B)

- ❏ A tentative selection has been made corresponding to the type of experimental design which is to be utilized as the basis for the research study (checklist attachment C)

- ❏ An experimental design notification memorandum has been circulated for review, with all appropriate parties signing off on the dependent, treatment, blocked, and limited variables included in the study; and the tentative design selected

Figure 5.7 (*Continued*)

Checklist Attachment A.

A Preliminary Breakdown of Dependent Variables and
Criterion Measures for Inclusion in the Experimental Design

Primary Dependent Variable(s)	Criterion Measure(s)	Measurement Scale
1.	1.	
	2.	
	3.	
	4.	
	5.	
2.	1.	
	2.	
	3.	
	4.	
	5.	
3.	1.	
	2.	
	3.	
	4.	
	5.	

Figure 5.7 (*Continued*)

Checklist Attachment A.

A Preliminary Breakdown of Dependent Variables and Criterion Measures for Inclusion in the Experimental Design

Secondary Dependent Variable(s)	Criterion Measure(s)	Measurement Scale
1.	1.	
	2.	
	3.	
2.	1.	
	2.	
	3.	
3.	1.	
	2.	
	3.	

Figure 5.7 *(Continued)*

Checklist Attachment B.

Table for the Organization of Treatment, Block, and Limited Variables

Treatment Variable(s)	Method	Classification	Type	Number of Levels	Level Description
	I - Incorporated N - Nested	QN - Quantitative QL - Qualitative	F - Fixed R - Random		
1.					
2.					
3.					
4.					
5.					
6.					
7.					
8.					
9.					
10.					

Block Variables	Number and Value for Levels Blocked	Variables Limited	At Level
1.		1.	
2.		2.	
3.		3.	
4.		4.	

Figure 5.7 (*Continued*)

Checklist Attachment C.

A Checksheet for the Identification of the Selected Experimental Design

		Place a check (✓) next to
	Experimental Design	the design selected

1. Completely Randomized Design
2. Randomized Block Design
3. Latin Square design
4. Graeco-Latin Square design
5. Balanced incomplete block design
6. Youden Square balanced incomplete block design
7. Partially balanced incomplete block design
8. Completely randomized factorial design
9. Randomized block factorial design
10. Completely randomized nested design
11. Completely randomized partially tested nested design
12. Split-plot design
13. Randomized block completely confounded factorial design
14. Latin Square completely confounded factorial design
15. Completely randomized fractional factorial design
    a. Extreme screening design
    b. Orthogonal Array design (specify L ____)
    c. High resolution fractional factorial design
    d. Other
16. Randomized block fractional factorial design
17. Completely randomized analysis of covariance design
18. Randomized block analysis of covariance design

Figure 5.7 (*Continued*)

Chapter 6

Sampling Procedures and Considerations

The purpose of this chapter is to review and discuss those premises and procedures associated with the sampling processes necessary to conduct a valid experiment. This discussion will revolve around the following topics:

1. Types of sampling plans and methods
2. Misconceptions related to the topic of randomness
3. Sample size and the precision of the experiment

Types of Sampling Plans and Methods

By this point, the following observation should be unnecessary, but the reader should note that randomization is required for sound experimental design. Without the utilization of randomization in the selection, assignment, and ordering of experimental units, we do not need to worry about the efficiency of the experiment; there will be none. This is because without randomization, there is no valid basis on which to analyze the data resulting from such a study.

The topic of sampling, the selection of a subgroup of experimental units from a research population, is a bit more complex than most novice practitioners may be led to believe. There are two general categories of sampling plans, that is, the method or procedure by which the experimental units are drawn from the research population: nonprobabilistic sampling plans and probabilistic sampling plans.

Nonprobabilistic sampling plans

The first category of sampling plans, of little concern to the industrial researcher, is nonrandom, or nonprobabilistic sampling plans. These plans are typified by the case where each sample or subgroup of size n in the research population does not have an equal probability of occurrence. It is important to note that in these cases, we have, in essence, redefined the research and target population of interest. Authors vary on the breakdown of these sampling plans. The following classification is common:

Judgment or purposive sampling plans. Judgment or purposive sampling plans are those plans in which the researcher selects the subjects or experimental units he or she wishes to include in the study. The selection is based upon the assumption that the representativeness of the group is known to be related to some group or constituency. Generally, philosophical and agreement research studies utilize purposive sampling plans. Note that this approach demands that no inferences or generalizations are made to larger populations.

Quota sampling plans. Quota sampling plans are identical to purposive sampling plans, except that the samples or subgroups are drawn (versus randomly selected) from specific and predetermined categories. This type of sampling plan would be used, for example, when we wished to compare opinions and impressions associated with a new type of domestic automobile for people who did and did not currently own a foreign car. Focus groups falling into purposively selected specific categories also are included in this classification.

Accident sampling plans. Not specifically identified by many authors, *accident sampling plans* are those in which the researcher simply selects individual subjects or experimental units seemingly at random, but in reality purposively. This condition results in the misconception that "no plan" is a "random plan," without regard to the research or target population to which an inference is to be made.

For example, imagine that a sales representative wanted to gather data related to what potential customers of a particular product thought about a new design or package. Wandering out into a department store where the product is sold, he or she stops people coming into the associated department, and asks them about the product or package, and records the results.

A second example would be that of a quality technician who, wishing to know the effects of a new incoming consumable material (i.e., a new supply source), wanders out onto the floor at a "random time," and selects a sample of 25 experimental units for analysis.

Finally, consider a relatively classical example (Wallis and Roberts, 1962), an article reported in the *New York Times* on May 22, 1950, displayed in Fig. 6.1. Additional information not originally stated in the article includes the facts that May 21 was a Sunday and the George Washington Bridge is a major bridge into New York City, south of Rockland County. As with any other accident sampling plan, the sample or subgroup selected becomes the population to which all inferences may be extended. In this example, "Sunday drivers"

ROUTE 17 AUTO JAM 10 MILES LONG

CAUSED BY THRUWAY BOARD'S POLL.

Sterlington, N.Y., May 21

A traffic jam that stalled a double line of autos for ten miles north of here developed tonight during a poll being conducted by the New York State Thruway Commission on Route 17.

Because of the unusually heavy traffic brought out by the fine weather, as many as 40,000 vehicles were affected or will be before the poll is completed at 7 a.m. tomorrow, observers estimated.

The poll, decided upon to help determine the southern route of the proposed superhighway between New York City and Buffalo, started at 7 a.m. today in this little community two miles north of Sloatsburg in Rockland County and twenty-eight miles north of the George Washington Bridge.

At that time the State Police started stopping every fourth automobile on Route 17, one of the main arteries to up-state areas, while canvassers for the commission asked the drivers these three questions: "Where did you come from?" "Where are you going?" "How often do you make this trip?"

Figure 6.1 *New York Times* Article.

were polled for traffic patterns that may have been used to estimate weekday traffic into New York City!

Probabilistic sampling plans

The second category associated with sampling plans is referred to as *probabilistic*, or *random*, sampling plans. Table 6.1 presents a breakdown of the major types of sampling plans within this category.

Note that the final two plans listed are nonprobabilistic sampling plans that we have already discussed. We will employ this table to discuss the principles

TABLE 6.1 Sampling Chart.

Type of Sampling	Brief Description	Advantages	Disadvantages
A. Simple Random	Assign to each population member a unique number: select sample items by use of random numbers	1. Requires minimum knowledge of population in advance 2. Free of possible Classification errors 3. Easy to analyze data and compute errors	1. Does not make use of any knowledge researchers may have of population 2. Larger errors for same sample size than in stratified sampling
B. Systematic	Use natural ordering or order population; select random starting point between 1 and the nearest integer to the sampling ratio (N/n); select items at interval of nearest integer to sampling ratio	1. If population is ordered with respect to a pertinent property, there is a stratification effect, and hence a reduction in variability compared to A 2. Simplicity of drawing sample; easy to check	1. If sampling interval is related to a periodic ordering of the population, increased variability may be introduced 2. Estimates of error likely to be high where there is stratification effect
C. Multistage Random	Use a form of random sampling in each sampling stage where there are at least two stages	1. Sampling lists, identification, and numbering required only for members of sampling units selected in sample 2. If sampling units are geographically defined, cuts down field costs (e.g., travel)	1. Errors likely to be larger than in A or B for same sample size 2. Errors increase as number of sampling units selected decreases
D. Stratified			
1. Proportionate	Select from every sampling unit at other than last stage, a random sample proportionate to size of sampling unit	1. Assures representatives with respect to property which forms basis of classifying units; therefore, yields less variability than A or C 2. Decreases chance of failing to include members of population because of classification process 3. Characteristics of each stratum can be estimated, and hence comparisons can be made	1. Requires accurate information on proportion of population in each stratum, otherwise increases error 2. If stratified lists are not available, may be costly to prepare them; possibility of faulty classification and hence increase in variability
2. Optimum Allocation	Same as 1 except sample is proportionate to variability within strata as well as their size	1. Less variability for same sample size than 1	1. Requires knowledge of variability of pertinent characteristic within strata
3. Disproportionate	Same as 1 except that size of sample is not proportionate to size of sampling unit, but is determined by analytical considerations or convenience	More efficient than 1 for comparison of strata or where different errors are optimum for different strata	Less efficient than 1 for determining population characteristics; i.e., more variability for same sample size
E. Cluster	Select sampling units by some form of random sampling; ultimate units are groups. Select these at random and take a complete count of each	1. If clusters are geographically defined, yields lowest field costs 2. Requires listing only individuals in selected clusters 3. Characteristics of clusters as well as those of population can be estimated 4. Can be used for subsequent samples, since clusters, not individuals, are selected, and substitution of individuals may be permissible	1. Larger errors for comparable size than other probability samples 2. Requires ability to assign each member of population uniquely to a cluster; inability to do so may result in duplication or omission of individuals
F. Stratified Cluster	Select clusters at random from every sampling unit	Reduces variability of plain cluster sampling	1. Disadvantages of stratified sampling added to those of cluster sampling 2. Since cluster properties may change, advantage of stratification may be reduced and make sample unusable for later research

TABLE 6.1 Sampling Chart. (*Continued*)

Type of Sampling	Brief Description	Advantages	Disadvantages
G. Repetitive: Multiple or Sequential	Two or more samples of any of the above types are taken, using results from earlier samples to design later ones, or determine if they are necessary	1. Provides estimates of population characteristics that facilitate efficient planning of succeeding sample, therefore, reduces error of final estimate 2. In the long run reduces number of observations required	1. Complicates administration of fieldwork 2. More computation and analysis required than in nonrepetitive sampling 3. Sequential sampling can only be used where a very small sample can approximate representativeness and where the number of observations can be increased conveniently at any stage of the research
H. Judgment	Select a subgroup of the population which, on the basis of available information, can be judged to be representative of the total population; take a complete count or subsample of this group	1. Reduces cost of preparing sample and field work, since ultimate units can be selected so that they are close together	1. Variability and bias of estimates cannot be measured or controlled 2. Requires strong assumptions or considerable knowledge of population and subgroup selected
I. Quota	Classify population by pertinent properties; determine desired proportion of sample from each class; fix quotas for each observer	1. Same as judgment 2. Introduces some stratification effect	1. Introduces bias of observers' classification of subjects and nonrandom selection within classes

From *The Design of Social Research*, by Russell L. Ackoff. Copyright© 1953. University of Chicago Press, Chicago. By permission of the publisher.

of random sampling. Note that, to a major degree, the selection of the random sampling plan to be employed is dictated by the structure of the experimental design constructed. Additionally, note that:

1. All random sampling plans are not simple random samples.
2. A random sample is not one where each individual experimental unit in the research population has an equal probability of selection. It is one where each sample of size n has an equal probability of selection.
3. Random sampling alone is not sufficient to conduct a valid experiment. Rather, random sampling or selection must be followed by a random assignment of the units to the treatments, and a random ordering of the application of the treatments and analytical procedures to guard against confounding the treatments with extraneous variable effects.

Misconceptions Related to the Topic of Randomness

Randomness is a term generally utilized to imply that we have selected, assigned, or ordered in such a way that the laws of probability may be applied. As pointed out by Wallis and Roberts (1962), there are a number of misconceptions associated with the concept of randomness in general and random sampling in particular. These misconceptions are found frequently enough, and are serious enough, to warrant a brief discussion.

1. *Random* is not to be confused with *haphazard, hit-or-miss,* or *aimless.* Randomness, as in random samples, must be attained through the use of some mechanical or electronic process that has been tested and verified. Haphazard arrangements do not often, except by chance, lead to levels of randomness adequate for our purposes. Test this premise by asking 100 people to randomly order the digits 1 through 10. Will any of the individuals present you with "1, 2, 3, 4, 5, 6, 7, 8, 9, 10?" The computer, in some sequence of the digits 1 to 10, will. People probably will not.

2. Randomness is not easy to achieve, and the researcher must take care to achieve a true random condition. Consider the following report:

> A classic illustration of inadequate randomization on a grand scale is the lottery used in establishing order numbers for the draft in 1940. Ten thousand numbers were written on slips of paper, the slips were put in capsules, and the capsules were put into a bowl and mixed. The capsules were then drawn by various blindfolded dignitaries in a public ceremony. The results showed marked departures from randomness. Apparently the difficulties of adequate mixing were not understood (Stouffer and Bartky, 1940).

3. The most common misconception shared by novice practitioners is that if we draw a true random sample, then that sample will be representative of the (research) population. This is simply not true. No sampling method can guarantee that the sample will be representative of the research population from which it was drawn. In a finite population of 100 crystal vases, for example, suppose that there are 5 vases with flaws. It is unlikely, but it is possible to randomly select 3 (again, for example) of the 5 vases as part of a sample of size 10. Obviously, 3 out of 10 defective units is not technically representative of the population of units. We draw random samples to maximize the probability that this will not occur, not to guarantee it.

4. It is not clear where this next misconception originates, but there are those individuals who believe that a lack of randomness in large samples is less important than in small ones. Randomness is important in any size sample. All of our inferential statistical tools and design premises are based upon the fundamental assumption that we have random samples, randomly applied to our treatments. Sample size does not change this premise. In fact, the assumption is actually reversed! Differences due to nonrandom influences would likely be assigned to sampling error in small sample differences. In large samples, where the experimental error term is driven down, differences due to nonrandom effects would be less likely to be attributed to sampling error.

Sample Size and the Precision of the Experiment

Optimum sample size (or the number of replications) is not a matter of opinion; it is a calculation based on variability, effect size, and risk. Amazingly, statisticians are often asked one of two questions regarding this issue:

1. "What size sample should I use?" Typically, this question is asked with no background information about the study to be designed. This question implies that there are "good" sample sizes, and "bad" sample sizes, and that the statistician or researcher can share the magic number. This would be amusing if it were not so often answered (usually with the answer "25" or "30" for no apparent reason).
2. "What proportion/percentage of the population should I draw for a sample?" Another startling question that must be rooted in the assumption that a particular percentage of a population must be studied to yield satisfactory or reliable results. As we have previously reviewed, this assumption is completely incorrect. The experimental error estimate is absolutely independent of the proportion of the (infinite) population studied.

The "bottom line" here is that the appropriate sample size is one that balances cost against risk. The calculation of an appropriate sample size (for continuous data) is dependent upon the following factors:

1. The effect size we are interested in detecting (Δ)
2. The variability of the populations in the experiment
3. The maximum risk the researcher is willing to accept that a Type I error will be committed in making an inference
4. The maximum risk the researcher is willing to accept that a Type II error will be committed in making an inference
5. The number of levels or treatment combinations that must be tested

Formulas for the calculation of sample sizes for one or two groups as related to means analyses are presented in *Experimental Design and Industrial Statistics—Level II* by Luftig (1991). Tables for sample sizes for various analyses have been published in many journals. Table 6.2 is presented as an example of how these tables are designed and used.

Let us review an example of the use of this table. Suppose we were conducting an experiment based upon a 2 × 3 full factorial design. Suppose further that the following conditions were stated:

1. Type I error is to be held at a maximum of 5 percent ($\alpha = .05$).
2. Type II error is to be held at a maximum of 10 percent ($\beta = .10$).
3. The effect size for the shift in means has been set at 2.00 units ($\Delta = 2.00$).
4. The level variances are assumed (based upon previous research) to be homogeneous, and equal to approximately 1.75 units ($\sigma^2 = 1.75$ and $\sigma = 1.33$).
5. Given that a 2 × 3 full factorial design is to be employed, there are six treatment combinations to be tested.

Referring to Table 6.2, we can see that for this example, with $\Delta/\sigma = 1.5$, $\alpha = .05$, and six treatment combinations, we would require 16 observations per

TABLE 6.2 Minimum Sample Size Needed to Ensure Power (1 − ß) = .90.

Δ / σ

No. of Levels	1.0 α				1.25 α				1.5 α				1.75 α				2.0 α				2.5 α				3.0 α			
	.2	.1	.05	.01	.2	.1	.05	.01	.2	.1	.05	.01	.2	.1	.05	.01	.2	.1	.05	.01	.2	.1	.05	.01	.2	.1	.05	.01
2	14	18	23	32	9	12	15	21	7	9	11	15	5	7	8	12	4	6	7	10	3	4	5	7	3	3	4	6
3	17	22	27	37	11	15	18	24	8	11	13	18	6	8	10	13	5	7	8	11	4	5	6	8	3	4	5	6
4	20	25	30	40	13	16	20	27	9	12	14	19	7	9	11	15	6	7	9	12	4	5	6	8	3	4	5	6
5	21	27	32	43	14	18	21	28	10	13	15	20	8	10	12	15	6	8	9	12	4	5	6	9	4	4	5	7
6	22	29	34	46	15	19	23	30	11	14	16	21	8	10	12	16	7	8	10	13	5	6	7	9	4	4	5	7
7	24	31	36	48	16	20	24	31	11	14	17	22	9	11	13	17	7	9	10	13	5	6	7	9	4	5	5	7
8	26	32	38	50	17	21	25	33	12	15	18	23	9	11	13	17	7	9	11	14	5	6	7	9	4	5	6	7
9	27	33	40	52	17	22	26	34	13	16	19	24	9	12	14	18	8	9	11	14	5	6	8	10	4	5	6	7
10	28	35	41	54	18	23	27	35	13	16	20	25	10	12	14	19	8	10	11	15	5	7	8	10	4	5	6	7
11	29	36	42	55	19	23	28	36	13	17	21	26	10	13	15	19	8	10	12	15	6	7	8	10	4	5	6	8
13	31	38	45	59	20	25	29	38	14	18	22	27	11	13	16	20	9	11	12	16	6	7	8	11	4	5	6	8
16	33	41	48	•	22	27	31	•	15	19	24	•	12	14	17	•	9	11	13	•	6	8	9	•	5	6	7	•
21	37	46	53	•	24	30	35	•	17	21	26	•	13	16	18	•	10	12	14	•	7	8	10	•	5	6	7	•
25	40	49	57	•	26	32	37	•	18	22	28	•	14	17	19	•	11	13	15	•	7	9	10	•	5	7	7	•
31	43	53	62	•	28	34	40	•	20	24	•	•	15	18	21	•	12	14	16	•	8	9	11	•	6	7	8	•

• Values not computed
* or treatment combinations

©1997 American Society for Quality Control. Reprinted with permission.

treatment combination (or run or cell). This observation brings up an important point. Under these conditions, we would require a total of 6 × 16, or 96 experimental unit observations. Some novice practitioners would, without exposure to a table of this type, assume that if we were to add a third treatment at two levels, we would double the number of experimental units required under the same effect size and error restrictions. Untrue. We would double the number of treatment combinations to twelve ($2^2 \times 3^1$) from six ($2^1 \times 3^1$), but the number of experimental units would more than double. Refer to the table once again. Liberally speaking, we would need at least 21 experimental units per combination, for a total of 252 units. In summary, as the number of treatment combinations increase, with all other conditions held constant, the number of observations per level or cell will also increase.

A final note on sampling and the subject of the meaning of the term "replication." Suppose we determine that we require 10 observations per run, or treatment combination, for a fractional factorial design. One question we might ask would be "Should we set up for the test condition, run the process, and draw a random sample of size 10; or should we run the test condition 10 times, drawing a sample of size one each time?"

In some cases, theoretical and industrial statisticians might disagree on the answer to this question. Theoretically, the most common response is that to obtain a "true" estimate of experimental error, one would have to set up and run the treatment condition 10 times, sampling once during each instance. To the industrial researcher, this procedure might not represent "true" experimental error at all. If we are studying a production process to determine the common causes of variability associated with the model, then the ten experimental setup conditions would not represent "true" experimental error at all. In fact, one could argue that the experiment run in this manner would have little external validity, particularly for those processes that are set up infrequently. In fact, the experiment might be conducted so as to identify those nominal set points where the process would be constantly held. In such a case, the repeated setup conditions associated with the 10 repeated experimental conditions would consist of two sources of error: experimental error, i.e., the unexplained replication variability present within a treatment combination, and the error due to the repetition effect, i.e., the variability between runs associated with the deviations from the targeted set points in the experimental run (as well as those changes common to the process through time). In many cases, we would construct our experiment such that both sources of variation are present, but the repetition effect would be analyzed as a factor in the model, separate from the residual (replication) term. Refer to Box, Hunter, and Hunter (1978) for an example of this condition as associated with the advanced discussion of Latin Squares with repetitions. Anderson and McLean (1974) also present a number of designs that take these conditions into consideration.

Summary

In summary, the replication of an experimental condition must be structured so as to provide an opportunity to calculate an experimental error term consistent with the target population and nature of the process studied. That is, the sampling plan must possess external validity.

Chapter

7

Establishing the Validity of the Data

The entire success of a scientific investigation depends upon the validity of the data generated through the study (Anderson and McLean, 1974). The term *validity* refers to the generic correctness of the data. But, the term also has a narrower, more specific meaning as applied to data collection instruments. It is of some surprise to researchers that novice practitioners will spend extensive amounts of time and energy developing complex designs for the manipulation of treatment variables and yet spend only a few minutes creating or using instruments of unknown precision and accuracy to measure the dependent variable (Spector, 1982).

Categories of Data

The first issue in the analysis of measurement validity is to recognize that the nature of the data collected will determine the indices and methods utilized to establish instrument validity. By *instrument*, we mean the device (e.g., survey), system (e.g., panel of raters), or equipment (e.g., micrometer) used to acquire data. Regardless of the instrument employed, all data are gathered within the context of a measurement process that must be shown to be operating in a state of statistical control.

Three categories of data are generally gathered: descriptive, criterion, and control data. The term *descriptive data* relates to all data gathered for studies associated with descriptive research. In the context of experimental, quasi-experimental, and relational research, descriptive data are those data that define the context or environment within which the study is to be conducted. *Criterion data* are those data gathered in order to test hypotheses or answer research questions. This category of data corresponds to studies that are almost always beyond descriptive in nature. Many authors describe a third

TABLE 7.1 A Breakdown of Measurement Scales and Data as Measured.*

Data Category	Measurement Scale		Description
Qualitative (Discrete, Attribute)	Nominal		Simple classification data, the assignment of numbers for identification purposes. Categorical scales.
	Ordinal	Count	Data corresponding to a frequency measure, such as number of pinholes or scratches per unit.
		Scale	Subjective scale data gathered from measurement or evaluative scales requiring the assignment of a numerical value; such as Likert scales and Semantic Differential scales.
		Low Resolution Continuous	Data resulting from low measurement resolution as related to the variability of the process or population measured.
Quantitative (Continuous, Variables)	Interval		Classification according to some continuum characterized by an equality of units, and the ability to sensibly subdivide the units of measure into smaller units.
	Ratio		Interval data, where the ratio between two measures on the scale is meaningful.

* The distinction is made between the nature of the underlying population and the data "delivered" by the instrument.

category of data referred to as *control data*. This category would relate to data gathered in order to control or block out extraneous variable effects. For example, control data would apply to block effects and covariate effects associated with the analysis of covariance (ANCOVA). The type of data gathered will also determine the nature of the approach and indices employed to establish the effectiveness of the instrumentation. Data may be portrayed and classified according to the matrix shown in Table 7.1.

The first step in properly assessing the effectiveness of the measurement process, and the correct statistical analysis of the data, is to properly categorize the type of data and measurement scale employed. Note that Table 7.1 presents this breakdown as associated with the data as delivered by the measurement process. Thickness, for example, may be described as an underlying continuous function. Provide an operator with a go–no go gauge, however, and we artificially reduce the measurement scale to a nominal level. It is absolutely essential that the practitioner, or statistician, be capable of determining the nature of the measurement scale and the nature of the underlying function describing the criterion measure.

Effectiveness of the Instrumentation

Once the nature of the criterion data is determined, the effectiveness of the instrumentation may be assessed. Table 7.2 provides an outline of the indices, metrics, and methods associated with this objective.

TABLE 7.2 Elements and Methods Associated with the Assessment of Instrumental Effectiveness.

	Elements			
	Precision	Accuracy	Reliability	Validity
Indices/Metrics	Measurement error	Bias	Agreement	Concordance with a standard
	due to	due to	due to	based on
	• Repeatability • Reproducibility	Out of calibration conditions	• Internal consistency • Concordance based on • Stability • Equivalence • Internal consistency or homogeneity	• Content Validity • Predictive Validity • Concurrent Validity • Construct Validity • Face Validity

Methods	• Repeatability and Reproducibility studies
	• Short-term control charts
	• Long-term control charts
		• Short-term calibration analyses
		• Long-term control charts
			• Test-retest methods
			• Equivalent forms analyses
			• Split-half and odd-even techniques; Kuder-Richardson analyses
				• Correlative analyses between the measurements obtained and:
				  • Subsequent scores or values
				  • Nonequivalent predicted score or value
				  • Metrics obtained through agreement analyses
				  • Assessments provided by targeted sample

	Quantitative Data		Qualitative Data	

Industrial applications of the assessment of measurement processes for control and capability where continuous data are involved are relatively well known or at least documented. For a comprehensive set of guidelines related to this topic, the practitioner is referred to *Guidelines for a Practical Approach to Gauge Capability Analysis* by Luftig (1991).

The analysis of measurement processes yielding qualitative data is not nearly as well documented in industry. Instead of referring to precision and accuracy, we are concerned with (respectively) reliability and validity. While the analysis of measurement systems yielding discrete data is beyond the scope of this text, some definitions of terms are in order at this point.

Reliability

Reliability, as associated with a measurement instrument, refers to a measure of consistency. There are three recognized reliability indices or coefficients:

1. *Stability:* Defined as the extent to which subjects or experimental units retain their relative ranks on repeated administrations of the same instrument.

2. *Equivalence:* Defined as the extent to which scores or observations obtained for the same subject or experimental unit correlate on equivalent or parallel analyses or forms.
3. *Internal consistency or homogeneity:* Defined as the extent to which the instrument or form is internally consistent. Methods of determining this condition include split-half and odd-even correlative techniques and special-purpose statistical tests such as the Kuder-Richardson index.

Factors affecting the reliability of a qualitative measurement instrument or device include (1) group or level variability, (2) the number of items on the form, or instrument length, and (3) the duration of the test-retest interval employed.

Validity

Validity is used as a generic term that relates to those methods used to determine the extent to which the scores, values, or observations obtained by an instrument correlate with some criterion external to the instrument itself. In essence, validity addresses whether the instrument measures what it is intended to measure, which is another way of describing the accuracy of the device. Validity differs from reliability in that reliability indices always reflect the correlation of the instrument with itself, where validity is assessed as the correlation between the instrument and some external measure. There are five generally recognized indices or coefficients of instrument validity:

1. *Content validity* refers to the assessment of accuracy through the analysis of one or more experts in determining whether the instrument actually measures its ostensible objectives.
2. *Predictive validity* is statistically measured. It is a function of whether an association exists between the scores, values, or observations obtained on an instrument and the subsequent (i.e., future) measure on the basis of a predetermined measure for the same subject or experimental unit. For example, much research has been conducted to determine if a score on the Scholastic Aptitude Test (SAT) has validity for the prediction of success in college (unfortunately, the answer is basically, no).
3. *Concurrent validity* is the same as predictive validity, except that a concurrent measure rather than a future metric is employed. The difference between this coefficient and the reliability coefficient of equivalence is that in the case of the reliability assessment, the association is established to an equivalent instrument or device. In the case of concurrent validity, the index is established as related to a nonequivalent criterion measure. An example of this type of index would be where an associative coefficient might be calculated for sampled critical customers' feedback on a 10-point scale related to the supplier's overall quality level and the amount of a particular commodity purchased from that supplier as a proportion of all other

customers. This associative (statistical) index could provide a measure of the concurrent validity related to the customer feedback form as a concurrent measure of purchasing tendencies. This example allows us to underscore a basic premise associated with the topic of reliability and validity. Specifically, an instrument or device may be reliable and invalid, but it cannot be valid and unreliable.

4. *Construct validity,* perhaps the most elusive of the coefficients of validity, is a method whereby the instrument or device is judged to be consistent with certain underlying theories or constructs. This coefficient generally does not play a major role in the assessment of validity within the industrial domain.

5. *Face validity,* simply stated, is the degree to which the subjects who are assessed by an instrument or device believe that the instrument is valid. This is generally not a statistically based index.

A great number of industrial personnel are unaware that the same premises of precision and accuracy associated with continuous data measurement processes may be extended to discrete data measurement systems. While it is true that the metrics and statistical tests are different, the principles are in fact exactly the same. For guidelines on the methods by which reliable and valid surveys and interview systems may be developed and assessed, refer to Guilford (1974).

For information on assessing ordinal scale and nominal data-based measurement processes for control and capability, refer to *Guidelines for a Practical Approach to the Assessment of Discrete Data Measurement Systems,* Vol. 1, Luftig (1993).

Regardless of the type of data involved and the assessment methodology employed, the instrument or device used to obtain criterion data must meet three requirements before the experiment may proceed:

1. The instrument must be precise or reliable.
2. The instrument must be accurate or valid.
3. The instrument must be operating in a state of statistical control.

As part of the design of any experiment, records should be completed and checked in order to determine that this fundamental analysis has been conducted. Figures 7.1 and 7.2 present sample forms that may be used for this purpose.

Updating the Planning Checklist for Industrial Research

Chapters 1 to 7 comprise the *plan* phase of our research design process and provide us with additional material with which to update the planning checklist we have been developing throughout this text. (See Fig. 7.3.).

Criterion Measure	Unit Of Measure	Instrument	Status Of The Measurement Process			
			Short-Term		Long-Term	
			In Control (Y/N)	Precision/ Tolerance Ratio	In Control (Y/N)	Precision/ Tolerance Ratio
1.						
2.						
3.						
4.						
5.						
6.						
7.						
8.						
9.						
10.						

Figure 7.1 Continuous Data Measurement Process Record and Status Form.

Criterion Measure	Unit Of Measure	Instrument	Scale *				Status Of The Measurement Process			
							Short-Term		Long-Term	
			1	2	3	4	In Control (Y/N)	Precision/ Tolerance Ratio	In Control (Y/N)	Precision/ Tolerance Ratio
1.										
2.										
3.										
4.										
5.										
6.										
7.										
8.										
9.										
10.										

* 1 = Nominal; 2 = Count; 3 = Severity or Scale; 4 = Weighted Count

Figure 7.2 Discrete Data Measurement Process Record and Status Form.

I. **Developing the Statement of the Problem**

- A. A comprehensive and complete background data base for the research study exists, originating from a: (✓ one)
  - ☐ Strategic product-market analysis
  - ☐ Technical competitive benchmarking study
  - ☐ Quality improvement effort
  - ☐ Problem-solving effort
  - ☐ Other (specify)_____

- B. The background data base includes a statement of significance for the problem

- C. The originating problem has been properly delineated and delimited

- D. A specific and focused statement of the problem has been developed that is: (all should have a ✓)
  - ☐ Concise
  - ☐ Precise
  - ☐ Testable
  - ☐ Obtainable

II. **Classifying the Research Study**

The research design to be conducted has been identified as: (✓ one)

- ☐ Agreement research
  - ○ Consensus
  - ○ Instrument/system concordance

- ☐ Descriptive research
  - ○ Status study
  - ○ Longitudinal study
  - ○ Case study
  - ○ Cross-sectional study

- ☐ Relational research
  - ○ Concurrent correlational
  - ○ Predictive correlational
  - ○ Causal comparative

- ☐ Experimental research

III. **Operationalizing the Statement of the Problem**

- ☐ If the research design is nonexperimental, research questions have been developed; or if the research design is experimental, research hypotheses have been developed

- ☐ The research questions or hypotheses have been evaluated and assessed as: (all should have a ✓)
  - ☐ Clear
  - ☐ Testable
  - ☐ Specific
  - ☐ Simply stated
  - ☐ Consistent with known facts

Figure 7.3 A Planning Checklist for Industrial Research.

IV. **Designing the Industrial Experiment** (all should have a ✓)

- ❏ The target and research populations have been described in clear and specific terms

- ❏ The primary dependent variable(s) and the associated criterion measure(s) has been identified, and is consistent with the research hypothesis(es) to be tested. Secondary dependent variables have also been identified. (The first two columns of checklist attachment A have been completed)

- ❏ All independent variables that may have an affect on the primary or secondary dependent variables have been identified

- ❏ The treatment variables have been identified and classified to be incorporated or nested

- ❏ The known and manipulable nuisance variables have been identified and classified to be blocked or limited

- ❏ The known and nonmanipulable nuisance variables have been identified and an appropriate monitoring plan has been set up for each factor for use during the conduct of the experiment

- ❏ The treatment variables have been correctly categorized as:
  * Qualitative versus quantitative
  * Fixed versus random effects

- ❏ Classification of fixed versus random effects for each treatment variable has been shown to be consistent with the statement of the problem and the research hypothesis(es)

- ❏ For each treatment variable, an appropriate number of levels has been identified. Level selections are appropriately within the operating envelope of the system, set at an extreme level, and have been shown to be nonoverlapping

- ❏ A summary table for independent variable management has been completed (checklist attachment B)

- ❏ A tentative selection has been made corresponding to the type of experimental design which is to be utilized as the basis for the research study (checklist attachment C)

- ❏ An experimental design notification memorandum has been circulated for review, with all appropriate parties signing off on the dependent, treatment, blocked, and limited variables included in the study; and the tentative design selected

ATTACHMENTS FOLLOW

Figure 7.3 (*Continued*)

Checklist Attachment A.

**A Preliminary Breakdown of Dependent Variables and
Criterion Measures for Inclusion in the Experimental Design**

Primary Dependent Variable(s)	Criterion Measure(s)	Measurement Scale
1.	1.	
	2.	
	3.	
	4.	
	5.	
2.	1.	
	2.	
	3.	
	4.	
	5.	
3.	1.	
	2.	
	3.	
	4.	
	5.	

Figure 7.3 (*Continued*)

Checklist Attachment A.

**A Preliminary Breakdown of Dependent Variables and
Criterion Measures for Inclusion in the Experimental Design**

Secondary Dependent Variable(s)	Criterion Measure(s)	Measurement Scale
1.	1.	
	2.	
	3.	
2.	1.	
	2.	
	3.	
3.	1.	
	2.	
	3.	

Figure 7.3 (*Continued*)

Checklist Attachment B.

Table for the Organization of Treatment, Block, and Limited Variables

Treatment Variable(s)	Method	Classification	Type	Number of Levels	Level Description
	I - Incorporated N - Nested	QN - Quantitative QL - Qualitative	F - Fixed R - Random		
1.					
2.					
3.					
4.					
5.					
6.					
7.					
8.					
9.					
10.					

Block Variables	Number and Value for Levels Blocked	Variables Limited	At Level
1.		1.	
2.		2.	
3.		3.	
4.		4.	

Figure 7.3 (*Continued*)

Checklist Attachment C.

A Checksheet for the Identification of the Selected Experimental Design

	Experimental Design	Place a check (✓) next to the design selected
1.	Completely Randomized Design	
2.	Randomized Block Design	
3.	Latin Square design	
4.	Graeco-Latin Square design	
5.	Balanced incomplete block design	
6.	Youden Square balanced incomplete block design	
7.	Partially balanced incomplete block design	
8.	Completely randomized factorial design	
9.	Randomized block factorial design	
10.	Completely randomized nested design	
11.	Completely randomized partially tested nested design	
12.	Split-plot design	
13.	Randomized block completely confounded factorial design	
14.	Latin Square completely confounded factorial design	
15.	Completely randomized fractional factorial design	
	a. Extreme screening design	
	b. Orthogonal Array design (specify L ____)	
	c. High resolution fractional factorial design	
	d. Other	
16.	Randomized block fractional factorial design	
17.	Completely randomized analysis of covariance design	
18.	Randomized block analysis of covariance design	

Figure 7.3 (*Continued*)

Checklist Attachment D. A breakdown of measurement scales and data as measured.*

Data Category	Measurement Scale		Description
Qualitative (Discrete, Attribute)	Nominal		Simple classification data; the assignment of numbers for identification purposes. Categorical scales.
	Ordinal	Count	Data corresponding to a frequency measure; such as number of pinholes or scratches per unit.
		Scale	Subjective scale data gathered from measurement or evaluative scales requiring the assignment of a numerical value; such as Likert Scales and Semantic Differential Scales.
		Low Resolution Continuous	Data resulting from low measurement resolution as related to the variability of the process or population measured.
Quantitative (Continuous, Variable)	Interval		Classification according to some continuum characterized by an equality of units, and the ability to sensibly subdivide the units of measure into smaller units.
	Ratio		Interval data, where the ratio between two measures on the scale is meaningful.

* The distinction is made between the nature of the underlying population and the data "delivered" by the instrument.

Figure 7.3 (*Continued*)

Checklist Attachment E. Elements and methods associated with the assessment of instrumentation effectiveness.

	Elements			
	Precision	**Accuracy**	**Reliability**	**Validity**
Indices/Metrics	Measurement error due to • Repeatability • Reproducibility	Bias due to Out of calibration conditions	Agreement due to • Internal consistency • Concordance based on • Stability • Equivalence • Internal consistency or homogeneity	Concordance with a standard based on • Content Validity • Predictive Validity • Concurrent Validity • Construct Validity • Face Validity

Methods	• Repeatability and Reproducibility studies
	• Short-term control charts
	• Long-term control charts
		• Short term calibration analyses
		• Long-term control charts
			• Test-retest methods
			• Equivalent forms analyses
			• Split-half and odd-even techniques; Kuder-Richardson analyses
				• Correlative analyses between the measurements obtained and:
				  • Subsequent scores or values
				  • Nonequivalent predicted score or value
				  • Metrics obtained through agreement analyses
				  • Assessments provided by targeted sample

	Quantitative Data		**Qualitative Data**	

Figure 7.3 (*Continued*)

Chapter 8

Managing the Execution of the Experiment

Considerations and Responsibilities

One of the most certain paths to catastrophe is to properly and painfully design a complex experiment, and then fail to manage its execution. Time after time, inexplicable results and conclusions which cannot be replicated are a direct result of a lack of discipline in the actual conduct of the study. For some reason, individuals willing to argue for hours over the "best" fractional design matrix to employ seem to find their rigor vaporized when the unglamorous work of actually running and monitoring the study begins. Yet it is exactly at this juncture that the study can be "lost." In general, the practitioner must consider the following elements in managing the experiment:

1. The size and complexity of the experiment
2. The total cost associated with the conduct of the study, as well as the specific cost of each experimental unit drawn and tested
3. The level of training related to experimental design theory possessed by the personnel working in the experimental environment
4. The experience in conducting successful experiments possessed by the personnel working in the experimental environment
5. Whether the experiment was initiated internally or externally, with respect to the location or environment within which the study must be conducted

Generally speaking, if the study is complex and costly, if the training and experience of the personnel in the experimental environment are limited, and especially if the requirement for the study is externally driven, then a maximum amount of on-site management should be anticipated. In many cases,

	Tasks to be Accomplished						
Responsibility	Measurement Analysis		Training and Instruction	On-Site Monitoring	Sampling	Testing	Data Analysis
	Continuous	Discrete					
Experimental Design Team Designate							
Statistical Facilitator							
Process Engineer							
Manufacturing Engineer							
Design Engineer							
Technician							
Operator							
First Line Supervisor							
Maintenance Personnel							
Procurement Personnel							
Quality Engineer							

P = Primary Responsibility
S = Secondary Responsibility

Figure 8.1 Responsibility Assignment Form.

particularly if the experimental units are expensive, a pilot study should be conducted.

The first step in successfully managing an experiment is to formally define individual responsibilities related to the tasks that must be performed. Figure 8.1 presents an example of a responsibility matrix employed for this purpose.

The Engineering Log

The second step in properly managing the experiment is to prepare and employ a comprehensive engineering log. Used in conjunction with check sheets and control charts, this device can ensure that:

1. The necessary conditions are properly set up for the treatment effects during each run (some software packages such as ANOVA-TM will automatically generate these run sheets).
2. The proper conditions for blocked and limited variables are in place during each experimental run.
3. The extraneous, nuisance variables are properly monitored during the conduct of the experiment.
4. The data associated with the criterion measures are properly obtained and recorded.

Figure 8.2 presents a sample layout for an engineering log which encompasses these elements.

Engineering Log
Treatment Variables Run Sheet

Date _____   Page ___ of ___

Experimental Run Number _____   Run Reference Number _____

Sponsor _____   Team Leader _____

Factor Number	Label	Treatment Variables	Setting	
			Level	Explanation
1	A			
2	B			
3	C			
4	D			
5	E			
6	F			
7	G			
8	H			
9	I			
10	J			
11	K			
12	L			
13	M			
14	N			
15	O			
Blocked Variables		1. _____ @ _____ 2. _____ @ _____		
Limited Variables		1. _____ @ _____ 2. _____ @ _____		

Figure 8.2 Engineering Log.

158   Chapter Eight

**Engineering Log
Nuisance (Nonmanipulable) Variables
Monitoring Sheet**

Date _____   Page ___ of ___

Experiment _____

Sponsor _____   Team Leader _____

Variable Number	Nuisance Variables	Monitoring Information			
		Checksheet		Control Chart	
		Form Number	Number of Observations/Run	Chart Number	Sample Size (n)
1					
2					
3					
4					
5					
6					
7					
8					
9					
10					
11					
12					
13					
14					
15					
16					
17					
18					

Figure 8.2 (*Continued*)

Engineering Log
Criterion Measure Data Recording Sheet

Date _____   Page ___ of ___

Experiment _____

Experimental Run Number _____   Run Reference Number _____

Sponsor _____   Team Leader _____

| Criterion Measure _____ |
| Responsibility _____ |

Order*	Measure	Order	Measure	Order	Measure
1		7		13	
2		8		14	
3		9		15	
4		10		16	
5		11		17	
6		12		18	

| Criterion Measure _____ |
| Responsibility _____ |

Order	Measure	Order	Measure	Order	Measure
1		7		13	
2		8		14	
3		9		15	
4		10		16	
5		11		17	
6		12		18	

*Order corresponds to test order across runs

Figure 8.2 (*Continued*)

These forms will be added to the planning checklist at the end of Chap. 10. The utilization of an engineering log should not be underestimated. In a number of cases, only the presence of the data gathered through this type of device allows for the analysis of residual effects traceable to nonmanipulable variables.

Chapter 9

Designing the Plan for the Statistical Analysis of the Data

The scope of this text does not allow for even a cursory overview of statistical methods. Many authors, in fact, make marked distinctions between those who design research and those who statistically analyze data (Hicks, 1964). They suggest that the professional researcher and the professional statistician possess markedly different skills, and work best in concert. While such a dichotomy may not be inevitable or desirable, the purpose of this text is primarily to review design methodology, not statistical applications.

In spite of an inability to review this topic in depth based upon the constraints of this text, some principles corresponding to the selection of analytical methods are briefly discussed.

Measurement Scales and Test Selection

The selection of a statistical test is based initially on the nature of the data and measurement scale associated with the dependent and (in some cases) independent variables. Table 9.1 presents a general overview of the relationship between the measurement scale of the data, and the selection of appropriate descriptive and inferential statistics.

As shown by this table, the first decision that must be made following the classification of the data is whether parametric or nonparametric methods of statistical inference are to be employed. Almost all of the techniques displayed in Chaps. 3, 4, and 5 are parametric applications.

Following the selection of the category of tests to be employed based upon the underlying assumptions of the tests and the nature of the data involved, a decision must be made as to the nature of the comparison or analysis to be conducted. This decision is predominantly related to the hypotheses to be tested or the research questions to be answered.

TABLE 9.1 Four Levels of Measurement and the Statistics Appropriate to Each Level.

Scale	Defining Relations	Examples of Appropriate Statistics	Appropriate Statistical Tests
Nominal	1. Equivalence	Mode Frequency Contingency coefficient	Nonparametric test
Ordinal	1. Equivalence 2. Greater than	Median Percentile Spearman r_s Kendall T Kendall W	Nonparametric test
Interval	1. Equivalence 2. Greater than 3. Known ratio of any two intervals	Mean Standard deviation Person product-moment correlation Multiple product-moment correlation	Nonparametric and parametric tests
Ratio	1. Equivalence 2. Greater than 3. Known ratio of any two intervals 4. Known ratio of any two scale values	Geometric mean Coefficient of variation	Nonparametric and parametric tests

Reprinted with permission from Siegel and Castellan (1988).

Parametric Testing and Analysis

As an example of the type of decisions to be made in the statistical analysis of continuous data with parametric methods, let us review a previous problem:

Statement of the problem: The purpose of this study is to determine whether packaging our CookieSnack cookies in a new foil overlay package will resist humidity incursion as well as our current wax paper package. Humidity analysis will be assessed utilizing the currently standard ASTM-B123 Laboratory Test.

Research hypotheses:

1. The new foil overlay package proposed for use in packaging our CookieSnack cookies will allow significantly lower levels of humidity to penetrate the package compared to our current wax paper package.

2. The new foil overlay package proposed for use in packaging our CookieSnack cookies will provide significantly lower levels of variability from cookie to cookie in the humidity permitted to penetrate the package as compared to our current wax paper package.

Figure 9.1 presents an example of a flowchart to guide the conduct of the statistical analysis of these data. The flowchart has been presented as an example of the complexity and rigor inherent to selecting an appropriate statistical analysis.

Nonparametric Testing and Analysis

In many cases, the data scale employed will not allow for the use of parametric methods. For example, determining whether two processes yield the same level of scratches or voids will generally lend itself to the use of a nonparametric analysis. Table 9.2 presents a comparison of parametric and nonparametric methods.

Table 9.3 presents an outline of the major nonparametric tests available, based upon the measurement scale employed and the hypothesis to be tested.

Figure 9.2 presents a simplified decision tree for the selection of an appropriate nonparametric test. For additional information on nonparametric analyses, refer to Gibbons (1976) and Siegel and Castellan (1988).

Finally, the reader should note that in many instances, it is not a difference between two or more groups that we are interested in assessing. Instead, the research question is oriented toward the assessment of an association or relationship. This area of statistical theory is not very well understood in the industrial environment. Further, the selection of an appropriate index for the analysis of association is extremely complex and well beyond the scope of this text.

The Seven-Step Procedure for Hypothesis Testing

The seven-step procedure for hypothesis testing is a convention utilized to provide, in a consistent fashion, all of the essential details associated with testing a set of statistical hypotheses. This general procedure is intended to make certain that a specific set of considerations is detailed in the use of inferential statistics, no matter what statistical test is used.

Blank forms for the seven-step procedure for hypothesis testing are included in App. D. In order to illustrate the configuration and use of this procedure, we will employ an example of a completely randomized design previously presented in Chap. 3.

Statement of the problem: The purpose of this study is to determine whether packaging our CookieSnack cookies in a new foil overlay package will resist humidity incursion as well as our current wax paper package. Humidity analysis will be assessed utilizing the currently standard ASTM-B123 Laboratory Test.

Research hypothesis: The new foil overlay package proposed for use in packaging our CookieSnack cookies will allow significantly lower levels of humidity to penetrate the package as compared to our current wax paper package.

The data collected for this study appeared as

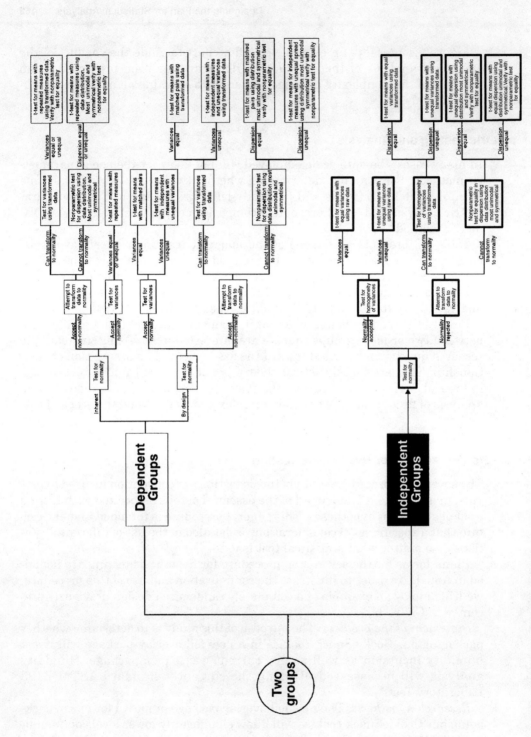

Figure 9.1 Flowchart for Selection of Appropriate Statistical Techniques for Comparing Two Groups.

TABLE 9.2 Analogy Between Nonparametric and Parametric Methods.

Type of Hypothesis	Name of Nonparametric Test(s)	Analogous Parametric Test(s)	Asymptomatic Relative Efficiency for Normal Distributions
Location parameter(s):			
One sample or paired sample	Sign test	Normal or Student's t-test	.637
One sample or paired sample	Wilcoxon signed rank test	Normal or Student's t-test	.955
Two independent samples	Mann-Whitney-Wilcoxon test	Normal or Student's t-test	.955
k independent samples	Kruskal-Wallis test	F test (one-way analysis of variance)	.955
k related samples	Friedman test	F test (randomized blocks analysis of variance)	$.955k/(k+1)$
Scale parameter(s):			
Two independent samples	Siegel-Tukey test	F test	.608
Association analysis:			
Two related samples	Speaman rank correlation or Kendall Tau	Pearson product-moment correlation	.912
k related samples	Kendall test	F test (randomized blocks analysis of variance)	$.955k/(k+1)$
	Kendall test	F test (balanced incomplete blocks analysis of variance)	$.955k/(k+1)$

Reprinted with permission from Gibbons (1976).

TABLE 9.3 Major Nonparametric Statistical Tests.*

Level of Measurement	One-Sample Case	Two-Sample Case		k-Sample Case		Nonparametric Measure of Correlation
		Dependent (Related) Samples	Independent Samples	Dependent (Related) Samples	Independent Samples	
Nominal	• Binomial test • χ^2 one-sample test	• McNemar test for the significance of changes	• Fisher exact probability test • χ^2 test for two independent samples	• Cochran Q test	• χ^2 test for k independent samples	• Contingency coefficient: C
Ordinal	• Kolmogorov-Smirnov one-sample test • One-sample runs test	• Sign test • Wilcoxon matched-pairs signed-ranks test	• Median test • Mann-Whitney U test • Kolmogorow-Smirnov two-sample test • Wald-Wolfowitz runs test • Moses test of extreme ractions Siegel-Tukey Test	• Friedman two-way analysis of variance by ranks	• Extension of the median test • Kruskal-Wallis one-way analysis of variance by ranks	• Spearman rank correlation coefficient: r_s • Kendall rank correlation coefficient: r • Kendall partial rank correlation coefficient: r_{xyz} • Kendall coefficien of concordance: $ $
Interval		• Walsh test • Randomization test for matched pairs	• Randomization test for two independent samples			

*Each column lists, cumulatively downward, tests applicable to the given level of measurement. For example, in the case of k related samples, when ordinal measurement has been achieved, both the Friedman two-way analysis of variance and the Cochran Q test may be applicable.

Reprinted with permission from Siegel and Castellan (1988).

166 Chapter Nine

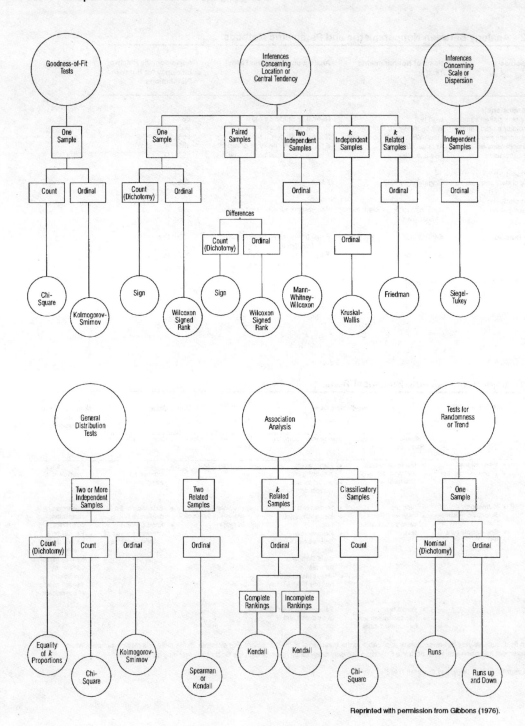

Figure 9.2 General Structure for the Selection of Nonparametric Tests.

Packaging Method (Treatment)

Foil overlay	Wax paper (control)
1.25	1.34
1.18	1.42
1.21	1.37
1.24	1.41
1.19	1.40
1.20	1.39
1.20	1.46
1.19	1.44
1.18	1.42
1.20	1.43

The descriptive statistics generated using SPSS for the two treatment levels are presented in Fig. 9.3.

We will use this research problem, and its associated data, to present an illustration of the seven-step procedure:

Step I: State the null (H_0) and research (H_1) hypotheses

For our example, these hypotheses would likely appear as:

$$H_0 : \mu_1 = \mu_2$$

$$H_1 : \mu_1 \neq \mu_2$$

Step II: State the maximum risk of committing a Type I error previously selected

The calculation of the sample size employed for the study would normally require that the researcher/experimental design team consider the following elements:

SPSS for MS WINDOWS Release 6.1

Number of valid observations (listwise) =   10.00

Variable	Mean	Std Dev	Valid Minimum	Maximum	N	Label
FOIL	1.20	.02	1.18	1.25	10	Foil Overlay
WAX	1.41	.03	1.34	1.46	10	Wax Paper (Control)

Figure 9.3 Descriptive Statistics for Humidity Levels for Packaging Experiment.

- The maximum risk for the commission of a Type I error
- The maximum risk for the commission of a Type II error
- An estimate of the variability inherent to the populations under investigation, which may also be expressed as the experimental error anticipated for the model
- The (minimum) effect size that the experiment is designed to detect.

All of these issues would have been explored prior to the collection of the data as associated with the experiment. The seven-step procedure would, therefore, be executed and presented well into the research design process. The reason that, as part of Step II, only the Type I error level is restated at this point is that (1) all of the conditions listed above would have been previously stated and employed in the calculation of the sample size (and therefore detailed in that section of the report and executive summary) and (2) only the Type I error level is required for the test of the null hypothesis.

For this example, this step might appear as

$$\alpha = .05$$

Step III: State the associated test statistic

The appropriate test statistic for the hypothesis test (as discussed earlier in this chapter) depends upon a number of factors such as the nature of the data, the condition of the levels to be compared (i.e., independent versus dependent), and the number of the treatment conditions involved in the analysis. This statement would appear as

$$\text{Test statistic} = \frac{\text{observed difference between the sample statistic and the hypothesized parameter}}{\text{estimated experimental error term}}$$

Step IV: Identify the random sampling distribution (RSD) of the test statistic when H_0 is true

In our example, the RSD statement would appear as "t is distributed according to the t distribution with $n_1 + n_2 - 2$ degrees of freedom."

Step V: State the critical value for rejecting the null hypothesis

For our example, the critical value for t would be obtained from a t table (App. C) based on $(10 + 10 - 2)$ 18 degrees of freedom and a two-tailed test at $\alpha = 0.05$, therefore:

$$\text{Reject } H_0 \text{ if ABS } (t) > 2.101$$

Step VI: Calculate the value of the test statistic from the sample data

For our example (refer to Fig. 3.7), the t value was 15.3 (on an absolute basis).

Step VII: Analyze the results and make an appropriate decision related to the null hypothesis

For this sample problem, we calculated the t and p values for these data using SPSS (Fig. 3.7). Based upon these calculations, our decision statement for this example might be presented as:

Given that $15.3 > 2.101$ (and $p = .000 < .05$), we will reject the null hypothesis and accept the research hypothesis. The probability of obtaining the observed difference due to sampling error alone if no difference truly existed between the population means is .000. Finally, the mean humidity incursion level for the control group (wax paper) is greater than for the foil overlay group of experimental units.

For additional information on this testing procedure see Hicks (1964), Kirk (1968), or *Experimental Design and Industrial Statistics—Level II* by Luftig (1991).

Chapter 10

Reporting and Standardizing the Results of the Research Study

Reporting the Results of the Study

Once the statistical analysis of the data is completed, the hypotheses may be tested, the research questions answered, and action taken as related to the statement of the problem originally posed. Appropriate action items will, naturally, vary with the purpose of the study. We cannot sensibly provide a set list of specific actions to be taken for a study to determine if three suppliers' incoming materials vary, and use the same checklist with a study based on a fractional factorial design intended to identify special causes of variation affecting a production process.

There are, however, some specific steps that should be taken in a logical progression:

1. Test the statistical hypotheses.
2. Test the research hypotheses or answer the research questions.
3. Prepare a comprehensive report of the research study incorporating the significant outcomes and conclusions.

Figure 10.1 provides a checklist for a recommended structure for this report and will provide a basis for our discussion of recommended action plans.

- ☐ Provide an overview of the background of the problem, and the reason for the conduct of the study.
- ☐ Present the statement of the problem.
- ☐ Present the research questions or research hypotheses.
- ☐ Present the research design employed, including:
  - ☐ The target population.
  - ☐ The dependent variable(s) and criterion measures.
  - ☐ The management plan for the independent variables (use of the independent variable layout form is recommended for this purpose).
  - ☐ The experimental design employed, including:
    - ☐ The type of experimental design
    - ☐ The design layout or array (include a linear graph if an orthogonal array is utilized)
    - ☐ Outcome of the sample size calculations
    - ☐ The sampling plan employed
    - ☐ The statistical hypotheses to be tested
- ☐ Present the results of the statistical analysis of the data for each criterion measure.
- ☐ Present the results of the statistical hypothesis tests (the use of Seven-Step Procedure forms in Appendix D are recommended for this purpose).
- ☐ Present the results and conclusions associated with the tests of the research hypotheses or answer the research questions posed.
- ☐ Summarize the results of the study as associated with the statement of the problem. Indicate the target population for which the results may be inferred. Describe the populations and conditions for which the results of the study may not be extended.
- ☐ Present the results of the confirmation experiment, if applicable.
- ☐ Provide recommendations for:
  - ☐ Further or additional research (verification experiments, descriptive research, secondary full factorial analyses, Evolutionary Operation [EVOP]).
- ☐ Present the next step in the action plan associated with the background of the problem.
- ☐ Completed SDCA checklist, if applicable.

Figure 10.1 A Recommended Checklist for the Final Report Resulting from a Research Study.

Product design review	Process design review	Materials procurement review	Additional/new training requirements
☐ Changes required to product targets or specifications	☐ Changes required to targets for first and/or second order critical process variables	☐ Action required for new vendor qualification	☐ Modification to training program/system for SOPs required for (specify): _____
☐ Changes required to engineering drawings	☐ Changes required to in-process SOPs	☐ Action required for vendor review	
☐ Changes required to in-process product specifications	☐ Changes required for inspection procedures	☐ Action required for vendor disqualification	☐ Follow-up training required for new control procedures for (specify): _____
☐ Changes required to tooling drawings and/or requirements	☐ New modified test standards required	☐ Action required for changes in targets or specifications for purchased consumable supplies or incoming raw materials	
☐ Concordance analysis to customer requirements necessary	☐ Modifications to existing and/or new control charts required		

Circulation list: _____

Figure 10.2 Standardize–Do–Check–Act Checklist.

The Standardize–Do–Check–Act (SDCA) Process

One of the important elements noted on the report checklist (Fig. 10.1) is the use of a standardize–do–check–act form. The use of a mechanism of this type will allow us to make certain that the potential gains resulting from the conduct of the experiment will not be lost in the short- or long-term future.

A sample form used for this purpose is presented in Fig. 10.2.

The Final Checklist for the Planning of Industrial Research

We may summarize the material reviewed in the last three chapters by reviewing the updated (and now final) checklist for the planning of industrial research (Fig. 10.3).

I. Developing the Statement of the Problem

☐ A. A comprehensive and complete background data base for the research study exists, originating from a: (✓ one)

☐ Strategic product-market analysis
☐ Technical competitive benchmarking study
☐ Quality improvement effort
☐ Problem-solving effort
☐ Other (specify)_____

☐ B. The background data base includes a statement of significance for the problem

☐ C. The originating problem has been properly delineated and delimited

☐ D. A specific and focused statement of the problem has been developed that is: (all should have a ✓)

☐ Concise
☐ Precise
☐ Testable
☐ Obtainable

II. Classifying the Research Study

The research design to be conducted has been identified as: (✓ one)

☐ Agreement research
○ Consensus
○ Instrument/system concordance

☐ Descriptive research
○ Status study
○ Longitudinal study
○ Case study
○ Cross-sectional study

☐ Relational research
○ Concurrent correlational
○ Predictive correlational
○ Causal comparative

☐ Experimental research

III. Operationalizing the Statement of the Problem

☐ If the research design is nonexperimental, research questions have been developed; or if the research design is experimental, research hypotheses have been developed

☐ The research questions or hypotheses have been evaluated and assessed as: (all should have a ✓)
☐ Clear
☐ Testable
☐ Specific
☐ Simply stated
☐ Consistent with known facts

Figure 10.3 A Planning Checklist for Industrial Research.

IV. Designing the Industrial Experiment (all should have a ✓)

- ☐ The target and research populations have been described in clear and specific terms

- ☐ The primary dependent variable(s) and the associated criterion measure(s) has been identified, and is consistent with the research hypothesis(es) to be tested. Secondary dependent variables have also been identified. (The first two columns of checklist attachment A have been completed)

- ☐ All independent variables that may have an affect on the primary or secondary dependent variables have been identified

- ☐ The treatment variables have been identified and classified to be incorporated or nested

- ☐ The known and manipulable nuisance variables have been identified and classified to be blocked or limited

- ☐ The known and nonmanipulable nuisance variables have been identified and an appropriate monitoring plan has been set up for each factor for use during the conduct of the experiment

- ☐ The treatment variables have been correctly categorized as:
  * Qualitative versus quantitative
  * Fixed versus random effects

- ☐ Classification of fixed versus random effects for each treatment variable has been shown to be consistent with the statement of the problem and the research hypothesis(es)

- ☐ For each treatment variable, an appropriate number of levels has been identified. Level selections are appropriately within the operating envelope of the system, set at an extreme level, and have been shown to be nonoverlapping

- ☐ A summary table for independent variable management has been completed (checklist attachment B)

- ☐ A tentative selection has been made corresponding to the type of experimental design which is to be utilized as the basis for the research study (checklist attachment C)

- ☐ An experimental design notification memorandum has been circulated for review, with all appropriate parties signing off on the dependent, treatment, blocked, and limited variables included in the study; and the tentative design selected

ATTACHMENTS FOLLOW

Figure 10.3 (*Continued*)

Checklist Attachment A.

A Preliminary Breakdown of Dependent Variables and Criterion Measures for Inclusion in the Experimental Design

Primary Dependent Variable(s)	Criterion Measure(s)	Measurement Scale
1.	1.	
	2.	
	3.	
	4.	
	5.	
2.	1.	
	2.	
	3.	
	4.	
	5.	
3.	1.	
	2.	
	3.	
	4.	
	5.	

Figure 10.3 (*Continued*)

Checklist Attachment A.

A Preliminary Breakdown of Dependent Variables and
Criterion Measures for Inclusion in the Experimental Design

Secondary Dependent Variable(s)	Criterion Measure(s)	Measurement Scale
1.	1.	
	2.	
	3.	
2.	1.	
	2.	
	3.	
3.	1.	
	2.	
	3.	

Figure 10.3 (*Continued*)

Checklist Attachment B.

Table for the Organization of Treatment, Block, and Limited Variables

Treatment Variable(s)	Method	Classification	Type	Number of Levels	Level Description
	I - Incorporated N - Nested	QN - Quantitative QL - Qualitative	F - Fixed R - Random		
1.					
2.					
3.					
4.					
5.					
6.					
7.					
8.					
9.					
10.					

Block Variables	Number and Value for Levels Blocked	Variables Limited	At Level
1.		1.	
2.		2.	
3.		3.	
4.		4.	

Figure 10.3 (*Continued*)

Checklist Attachment C.

A Checksheet for the Identification of the Selected Experimental Design

	Experimental Design	Place a check (✓) next to the design selected
1.	Completely Randomized Design	
2.	Randomized Block Design	
3.	Latin Square design	
4.	Graeco-Latin Square design	
5.	Balanced incomplete block design	
6.	Youden Square balanced incomplete block design	
7.	Partially balanced incomplete block design	
8.	Completely randomized factorial design	
9.	Randomized block factorial design	
10.	Completely randomized nested design	
11.	Completely randomized partially tested nested design	
12.	Split-plot design	
13.	Randomized block completely confounded factorial design	
14.	Latin Square completely confounded factorial design	
15.	Completely randomized fractional factorial design	
	a. Extreme screening design	
	b. Orthogonal Array design (specify L ____)	
	c. High resolution fractional factorial design	
	d. Other	
16.	Randomized block fractional factorial design	
17.	Completely randomized analysis of covariance design	
18.	Randomized block analysis of covariance design	

Figure 10.3 (*Continued*)

Checklist Attachment D. A breakdown of measurement scales and data as measured.*

Data Category	Measurement Scale		Description
Qualitative (Discrete, Attribute)	Nominal		Simple classification data; the assignment of numbers for identification purposes. Categorical scales.
	Ordinal	Count	Data corresponding to a frequency measure; such as number of pinholes or scratches per unit.
		Scale	Subjective scale data gathered from measurement or evaluative scales requiring the assignment of a numerical value; such as Likert Scales and Semantic Differential Scales.
		Low Resolution Continuous	Data resulting from low measurement resolution as related to the variability of the process or population measured.
Quantitative (Continuous, Variable)	Interval		Classification according to some continuum characterized by an equality of units, and the ability to sensibly subdivide the units of measure into smaller units.
	Ratio		Interval data, where the ratio between two measures on the scale is meaningful.

* The distinction is made between the nature of the underlying population and the data "delivered" by the instrument.

Figure 10.3 (*Continued*)

Checklist Attachment E. Elements and methods associated with the assessment of instrumentation effectiveness.

	Elements			
	Precision	Accuracy	Reliability	Validity
Indices/Metrics	Measurement error due to • Repeatability • Reproducibility	Bias due to Out of calibration conditions	Agreement due to • Internal consistency • Concordance based on • Stability • Equivalence • Internal consistency or homogeneity	Concordance with a standard based on • Content Validity • Predictive Validity • Concurrent Validity • Construct Validity • Face Validity

Methods	• Repeatability and Reproducibility studies
	• Short-term control charts
	• Long-term control charts
		• Short term calibration analyses
		• Long-term control charts
			• Test-retest methods
			• Equivalent forms analyses
			• Split-half and odd-even techniques; Kuder-Richardson analyses
				• Correlative analyses between the measurements obtained and:
				  • Subsequent scores or values
				  • Nonequivalent predicted score or value
				  • Metrics obtained through agreement analyses
				  • Assessments provided by targeted sample

	Quantitative Data		Qualitative Data	

Figure 10.3 (*Continued*)

Checklist Attachment F. Responsibility assignment form.

Responsibility	Tasks to be Accomplished						
	Measurement Analysis		Training and Instruction	On-Site Monitoring	Sampling	Testing	Data Analysis
	Continuous	Discrete					
Experimental Design Team Designate							
Statistical Facilitator							
Process Engineer							
Manufacturing Engineer							
Design Engineer							
Technician							
Operator							
First Line Supervisor							
Maintenance Personnel							
Procurement Personnel							
Quality Engineer							

P = Primary Responsibility
S = Secondary Responsibility

Figure 10.3 (*Continued*)

Checklist Attachment G. Engineering log.

**Engineering Log
Treatment Variables Run Sheet**

Date _____   Page ___ of ___

Experimental Run Number _____   Run Reference Number _____

Sponsor _____   Team Leader _____

Factor #	Label	Treatment Variables	Setting	
			Level	Explanation
1	A			
2	B			
3	C			
4	D			
5	E			
6	F			
7	G			
8	H			
9	I			
10	J			
11	K			
12	L			
13	M			
14	N			
15	O			
Blocked Variables		1. _____	@	_____
		2. _____	@	_____
Limited Variables		1. _____	@	_____
		2. _____	@	_____

Figure 10.3 (*Continued*)

Checklist Attachment G continued.

Engineering Log
Nuisance (Nonmanipulable) Variables
Monitoring Sheet

Date _____   Page ___ of ___

Experiment _____

Sponsor _____   Team Leader _____

Variable Number	Nuisance Variables	Monitoring Information		
		Checksheet		Control Chart	
		Form Number	Number of Observations/ Run	Chart Number	Sample Size (n)
1					
2					
3					
4					
5					
6					
7					
8					
9					
10					
11					
12					
13					
14					
15					
16					
17					
18					

Figure 10.3 (*Continued*)

Checklist Attachment G continued.

Engineering Log
Criterion Measure Data Recording Sheet

Date _____   Page ___ of ___

Experiment _____

Experimental Run Number _____   Run Reference Number _____

Sponsor _____   Team Leader _____

Criterion Measure _____

Responsibility _____

Order*	Measure	Order	Measure	Order	Measure
1		7		13	
2		8		14	
3		9		15	
4		10		16	
5		11		17	
6		12		18	

Criterion Measure _____

Responsibility _____

Order	Measure	Order	Measure	Order	Measure
1		7		13	
2		8		14	
3		9		15	
4		10		16	
5		11		17	
6		12		18	

*Order corresponds to test order across runs

Figure 10.3 (*Continued*)

Checklist Attachment H.

A Recommended Checklist for the Final Report Resulting from a Research Study

☐ Provide an overview of the background of the problem, and the reason for the conduct of the study.

☐ Present the statement of the problem.

☐ Present the research questions or research hypotheses.

☐ Present the research design employed, including:

☐   The target population.

☐   The dependent variable(s) and criterion measures.

☐   The management plan for the independent variables (use of the independent variable layout form is recommended for this purpose).

☐   The experimental design employed, including:

☐   The type of experimental design

☐   The design layout or array (include a linear graph if an orthogonal array is utilized)

☐   Outcome of the sample size calculations

☐   The sampling plan employed

☐   The statistical hypotheses to be tested

☐ Present the results of the statistical analysis of the data for each criterion measure.

☐ Present the results of the statistical hypothesis tests (the use of Seven-Step Procedure forms in Appendix D are recommended for this purpose).

☐ Present the results and conclusions associated with the tests of the research hypotheses or answer the research questions posed.

☐ Summarize the results of the study as associated with the statement of the problem. Indicate the target population for which the results may be inferred. Describe the populations and conditions for which the results of the study may not be extended.

☐ Present the results of the confirmation experiment, if applicable.

☐ Provide recommendations for:

☐   Further or additional research (verification experiments, descriptive research, secondary full factorial analyses, Evolutionary Operation [EVOP]).

☐ Present the next step in the action plan associated with the background of the problem.

☐ Completed SDCA checklist, if applicable.

Figure 10.3 *(Continued)*

Checklist Attachment I. SDCA checklist.

Product design review	Process design review	Materials procurement review	Additional/new training requirements
☐ Changes required to product targets or specifications	☐ Changes required to targets for first and/or second order critical process variables	☐ Action required for new vendor qualification	☐ Modification to training program/system for SOPs required for (specify): _____ _____
☐ Changes required to engineering drawings	☐ Changes required to in-process SOPs	☐ Action required for vendor review	
☐ Changes required to in-process product specifications	☐ Changes required for inspection procedures	☐ Action required for vendor disqualification	☐ Follow-up training required for new control procedures for (specify): _____ _____
☐ Changes required to tooling drawings and/or requirements	☐ New modified test standards required	☐ Action required for changes in targets or specifications for purchased consumable supplies or incoming raw materials	
☐ Concordance analysis to customer requirements necessary	☐ Modifications to existing and/or new control charts required		

Circulation list: _____ _____
_____ _____

Figure 10.3 (*Continued*)

Chapter

11

Designing the Industrial Experiment: Case Studies

In this chapter, examples will be provided for some of the major forms of experimental designs identified earlier. Through these case studies, we will explore the logic and purpose of experimental design as outlined in previous chapters. Additionally, a review of these designs should enhance the reader's capability to select an appropriate design following the classification of treatment, block, and limited factors (refer to Fig. 5.5).

The planned structure for the review of each of these case studies will be presented consistently as follows:

1. A general description of the design under discussion will be provided. This will include the advantages and/or disadvantages of the design. Important considerations and underlying assumptions in the structure and use of the experimental design will also be reviewed.
2. The use of the design will be exemplified by the application of a sample experiment. A statement of the problem associated with the case study will be presented. This will be followed by:
   a. One or more research
   b. The statistical hypotheses to be tested
3. A sample table for the organization of the independent variables will be provided, followed by the design of the experiment itself.
4. An example of the analysis of some or all of the data collected will be provided.
5. Results and conclusions corresponding to the statistical and research hypotheses will be presented and/or developed through self-review activities. Sample answers and a summary of observations appropriate to the use and structure of the design will be presented in App. E.

Case Study 1: A Completely Randomized Design

The completely randomized design is the most basic of the true experimental designs. It is best employed in those cases where a large degree of control may be exercised over extraneous variables, and where the experimental units are relatively homogeneous (Kirk, 1968). The generally accepted advantages of this design may be expressed (Juran, Gryna, and Bingham, 1979) as

1. The researcher has maximum flexibility in terms of the number of treatment levels that may be tested, and the number of experimental units assigned to each level.
2. The design generally is not compromised in the event of lost or missing data (within reason).
3. The analysis of a completely randomized design is relatively straightforward.

We refer to this design as *randomized* because the experimental units are randomly sampled and randomly assigned to the levels of the treatment. The term *completely* is employed to differentiate this design from a randomized block design, utilized when the experimental units are not nearly so homogeneous.

In order to illustrate the structure of this design, we will use the following research problem:

Statement of the problem: The purpose of this experiment is to determine which of four coolant suppliers provides the highest quality incoming product for use on the no. 6 machine at the Albion Plant. The target population for this study consists of all material produced at the Albion Plant, with part quality identified as the dependent variable. The criterion measure for this study is the (coded) deviation from target.

Research hypotheses:

1. Vendor 4 (our current coolant supplier) provides coolant which, when used on the no. 6 machine under standard operating conditions, produces lower deviation from target product than would be produced under the same conditions with coolant supplied by any of the three new proposed suppliers.
2. There is no significant difference in the variability of deviation from target values for product manufactured with coolant supplied from our current or three new proposed suppliers.

Statistical hypotheses to be tested:

1. H_0: $\mu_1 = \mu_2 = \mu_3 = \mu_4$

   H_1: $\mu_1 \neq \mu_2 \neq \mu_3 \neq \mu_4$

   and

2. H_0: $\sigma_1^2 = \sigma_2^2 = \sigma_3^2 = \sigma_4^2$

   H_1: $\sigma_1^2 \neq \sigma_2^2 \neq \sigma_3^2 \neq \sigma_4^2$

   which requires that we also test

3. H_0: $\gamma_3 = 0$

   H_1: $\gamma_3 \neq 0$

   (for each supplier) and

4. H_0: $\gamma_4 = 0$

   H_1: $\gamma_4 \neq 0$

(for each supplier).

Statistical hypothesis 1 tests whether the mean deviation from target is the same for product produced with coolant from all four suppliers (H_0) or whether there is a statistically significant difference (H_1). If a difference exists, a mean (μ) that is statistically significantly closer to zero would be preferred, since that case would represent the least amount of deviation from target (on average); i.e., with that coolant supplier, material produced on the no. 6 machine at the Albion Plant comes closer to meeting the target for part quality on average than material produced with coolant from the other suppliers.

Statistical hypothesis 2 tests whether the variance is the same for all four suppliers (H_0) or whether there is a statistically significant difference (H_1). If a difference exists, a variance (σ^2) that is statistically significantly lower would be preferred, since that case would represent the least amount of variability; i.e., with that coolant supplier, material produced on the no. 6 machine at the Albion Plant has more consistent part quality than material produced with coolant from the other suppliers.

Statistical hypothesis 3 tests skewness (γ_3) for data from each supplier. If $\gamma_3 = 0$ (H_0), then the data are not skewed, data fall relatively symmetrically about the mean. If the data are skewed (H_1: $\gamma_3 \neq 0$), values are not symmetric. In the case of part quality, a γ_3 value closest to zero would be preferred. Another reason for testing skewness is for an assessment of normality. Data that can be approximated by a normal distribution will have skewness (γ_3) and kurtosis (γ_4) values close to zero. If the data are skewed, they are not normally distributed and that condition must be known in order to select the appropriate test of means and variances. If variances are equal, the one-way analysis of variance (ANOVA) is an appropriate test of means no matter the shape of the underlying distributions, but if variances are unequal, the shape of the distributions becomes a factor. Also, the test of variances depends on the assumption of normality (although the Levene test is more robust to departures from normality than the Bartlett-Box F test).

Treatment Variable(s)	Method I - Incorporated N - Nested	Classification QN - Quantitative QL - Qualitative	Type F - Fixed R - Random	Number of Levels	Level Description
1. *Coolant Supplier*	*I*	*QL*	*F*	*4*	*Supplier 1/2/3/4*
2.					
3.					
4.					
5.					
6.					
7.					
8.					
9.					
Blocked Variables	**Number and Value for Levels Blocked**		**Variables Limited**		**At Level**
1. *None*			1. *Plant*		*Albion*
2.			2. *Machine*		*No. 6 Machine*
3.			3.		
4.			4.		

Figure 11.1 Organization of Treatment, Block, and Limited Variables for Case Study 1.

Similarly, statistical hypothesis 4 tests kurtosis (γ_4) for data from each supplier. If $\gamma_4 = 0$ (H_0), then the data are mesokurtic, as with a normal distribution. If $\gamma_4 < 0$, the distribution is platykurtic, or "flat." If $\gamma_4 > 0$, the distribution is leptokurtic, or "peaked." In the case of part quality, a higher γ_4 value would be preferred since that would indicate a concentration of values at the center point (preferably at the target value). Again, it is important to determine whether the data are normally distributed in order to select the appropriate tests for means and variances.

The experiment currently under consideration has been organized on the basis of the data presented in Fig. 11.1.

The structure of the completely randomized experiment is relatively straightforward and can generally be displayed as

Treatment levels

$$\begin{array}{ccccc} j_1 & j_2 & j_3 & j_4 & \cdots \quad J \\ X_{11} & X_{12} & X_{13} & X_{14} & \\ X_{21} & X_{22} & X_{23} & X_{24} & \\ X_{31} & X_{32} & X_{33} & X_{34} & \\ X_{41} & X_{42} & X_{43} & X_{44} & \\ X_{51} & X_{52} & X_{53} & X_{54} & \\ X_{61} & X_{62} & X_{63} & X_{64} & \end{array}$$

In this experiment, there are four treatment levels (coolant suppliers) with each treatment condition replicated ten ($n = 10$) times. The data collected for the study appeared as

Treatment Levels (Supplier)			
1	2	3	4
10.64	11.69	18.77	10.41
10.38	7.63	19.21	11.31
10.61	12.61	18.30	11.72
10.82	10.96	20.16	14.12
12.73	11.56	13.95	10.63
15.21	14.63	22.18	11.09
16.76	12.67	18.17	12.81
14.94	9.07	19.85	9.29
13.79	14.05	18.17	11.37
8.51	10.17	16.00	11.29

The analytical strategy to be employed with this model consists of two primary analyses:

1. A one-way ANOVA to test the statistical hypotheses associated with the means
2. The Levene test, to test the statistical hypotheses associated with the variances.

Given that the test for variances possesses underlying assumptions associated with the shape of the population distributions from which the samples were randomly drawn, each treatment level will also be tested for normality. The selection of tests and inferential statistics employed for this problem is not arbitrary. The selection of the appropriate test statistic is based upon the nature of the data, the number of factors involved, the number of treatment levels, and the underlying assumptions that may or may not be met. The contribution of each of these elements and an outline of how these tests are selected were presented in Chap. 5.

For each of the analyses presented in this chapter, statisticians would generally describe the analysis in the form of a mathematical model. These models provide the breakdown of the components of variability, and guide the analytical procedure. The nature and structure of these models is beyond the scope of this book. For additional information on this topic, refer to *Experimental Design and Industrial Statistics—Level III* by Luftig (1991).

Step Number	Description	Participant Response
Step I	State the null and research hypotheses.	$H_0: \quad \mu_1 = \mu_2 = \mu_3 = \mu_4$ $H_1: \quad \mu_1 \neq \mu_2 \neq \mu_3 \neq \mu_4$
Step II	State the maximum risk of committing a Type I error.	$\alpha = 0.10$
Step III	State the associated test statistic.	$F = MS_T / MS_E$
Step IV	Identify the random sampling distribution of the test statistic when H_0 is true.	$F \underset{=}{d} F(3, 36)$ df when H_0 is true
Step V	State the critical value for rejecting the null hypothesis.	Reject H_0 if $p < .10$
Step VI	Calculate the value of the test statistic from the sample data.	$F =$
Step VII	Analyze the results and make an appropriate inference related to the hypotheses tested.	$p =$ Discussion:

Figure 11.2 Seven-step Procedure for Hypothesis Testing of Means.

Our primary tests are the tests of means (statistical hypothesis 1) and the test of variances (statistical hypothesis 2). Figures 11.2 and 11.3 illustrate the seven-step procedure for testing (introduced in Chap. 9). At this point we can complete the first five steps.

To ensure we are using the appropriate tests for means and variances, statistical hypotheses 3 (skewness) and 4 (kurtosis) must be tested first. These represent tests of normality. The software program Normal allows us to test for normality using the Anderson-Darling, Lin-Mudholkar, and moments tests. Tests of skewness and kurtosis (moments tests) are most appropriate for

Step Number	Description	Participant Response
Step I	State the null and research hypotheses.	$H_0:\ \sigma_1^2 = \sigma_2^2 = \sigma_3^2 = \sigma_4^2$
		$H_1:\ \sigma_1^2 \neq \sigma_2^2 \neq \sigma_3^2 \neq \sigma_4^2$
Step II	State the maximum risk of committing a Type I error.	$\alpha = 0.10$
Step III	State the associated test statistic.	Levene F
Step IV	Identify the random sampling distribution of the test statistic when H_0 is true.	$F \underset{=}{d} F\ (3, 36)$ df when H_0 is true
Step V	State the critical value for rejecting the null hypothesis.	Reject H_0 if $p < .10$
Step VI	Calculate the value of the test statistic from the sample data.	$F =$
Step VII	Analyze the results and make an appropriate inference related to the hypotheses tested.	$p =$ Discussion:

Figure 11.3 Seven-step Procedure for Hypothesis Testing of Variances.

sample sizes larger than 25. With a sample size of 10, the Anderson-Darling and the Lin-Mudholkar tests are more appropriate. The Anderson-Darling test is sensitive to small sample sizes and kurtosis. The Lin-Mudholkar test is sensitive to skewness. Results are presented in Fig. 11.4.

Since both tests give us p values greater than .10 (for all four suppliers), we can assume that the distributions (of deviation from target for parts made with coolant from each of the four coolant suppliers) can be approximated by a normal distribution.

The next step is to run the one-way ANOVA procedure in a standard statis-

Anderson-Darling

Col	n	Mean	Std Dev	Low	High	Range	$A^{2}*$	p
1	10	12.44	2.65	8.51	16.76	8.25	0.392	0.396
2	10	11.50	2.16	7.63	14.63	7.00	0.165	0.959
3	10	18.48	2.26	13.95	22.18	8.23	0.385	0.414
4	10	11.40	1.32	9.29	14.12	4.83	0.439	0.308

Lin-Mudholkar

Col	n	Mean	Std Dev	Low	High	Range	r	p
1	10	12.44	2.65	8.51	16.76	8.25	0.242	0.663
2	10	11.50	2.16	7.63	14.63	7.00	0.282	0.609
3	10	18.48	2.26	13.95	22.18	8.23	0.373	0.486
4	10	11.40	1.32	9.29	14.12	4.83	0.440	0.399

Box & Whisker

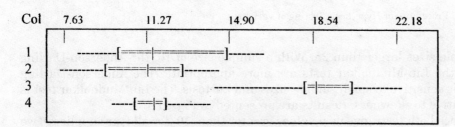

Figure 11.4 Tests for Normality Using Normal Software.

SPSS for MS WINDOWS Release 6.1

----- O N E W A Y -----

Variable DEVIATION Deviation from target
By Variable SUPPLIER

Analysis of Variance

Source	D.F.	Sum of Squares	Mean Squares	F Ratio	F Prob.
Between Groups	3	342.5509	114.1836	24.6402	.0000
Within Groups	36	166.8256	4.6340		
Total	39	509.3765			

Levene Test for Homogeneity of Variances

Statistic	df1	df2	2-tail Sig.
1.9421	3	36	.140

Figure 11.5 One-way ANOVA Using SPSS/PC Software to Assess Differences in Means and Variances for Deviation from Target Part Quality for Four Coolant Suppliers.

tical software package (SPSS/PC for Windows), which provides the analysis we require for both sets of the primary statistical hypotheses. Results of this analysis are displayed in Fig. 11.5.

The ANOVA table for the analysis which tests statistical hypothesis 1 (differences in mean deviation from target) reveals an F value of 24.6402, which corresponds to a p value of .0000. Since the p value is less than the significance level we selected ($\alpha = .10$), this indicates that there is a significant difference between the four coolant suppliers. (The null hypothesis, H_0, is rejected and the alternative hypothesis, H_1, is accepted.)

Steps 6 and 7 of the seven-step procedure for the test of means can now be added (Fig. 11.6). The Levene test is used to test statistical hypothesis 2 (differences in variances). SPSS calculates an F value for the Levene W statistic. The analysis reveals an F value of 1.9421, which corresponds to a p value of .140. Since the p value is not less than the significance level we selected ($\alpha = .10$), there is no statistical evidence to infer a difference between the four coolant suppliers in terms of variability. (The null hypothesis, H_0, is accepted.)

Steps 6 and 7 of the seven-step procedure for the test of variances can now be added (Fig. 11.7). Now, we know that there is a significant difference between the means of the four coolant suppliers, but no significant difference between the variances. The next questions to be addressed are: Do all of the

Step Number	Description	Participant Response
Step I	State the null and research hypotheses.	$H_0: \mu_1 = \mu_2 = \mu_3 = \mu_4$ $H_1: \mu_1 \neq \mu_2 \neq \mu_3 \neq \mu_4$
Step II	State the maximum risk of committing a Type I error.	$\alpha = 0.10$
Step III	State the associated test statistic.	$F = MS_T / MS_E$
Step IV	Identify the random sampling distribution of the test statistic when H_0 is true.	$F \underset{=}{d} F(3, 36)$ df when H_0 is true
Step V	State the critical value for rejecting the null hypothesis.	Reject H_0 if $p < .10$
Step VI	Calculate the value of the test statistic from the sample data.	$F = 24.6402$
Step VII	Analyze the results and make an appropriate inference related to the hypotheses tested.	$p = .0000$ Discussion: Reject H_0. We have sufficient statistical evidence to infer that there is a difference between the four coolant suppliers on mean deviation from target.

Figure 11.6 Seven-step Procedure for Hypothesis Testing of Means.

coolant suppliers have significantly different means or are some of them approximately equivalent? Which of the coolant suppliers is "best"?

The Student-Newman-Keuls procedure provides such an analysis (called a post-hoc test). It is displayed in Fig. 11.8, which shows that, while there is no significant differences in mean deviation from target for coolant suppliers 1, 2, and

Step Number	Description	Participant Response
Step I	State the null and research hypotheses.	$H_0: \quad \sigma_1^2 = \sigma_2^2 = \sigma_3^2 = \sigma_4^2$
		$H_1: \quad \sigma_1^2 \neq \sigma_2^2 \neq \sigma_3^2 \neq \sigma_4^2$
Step II	State the maximum risk of committing a Type I error.	$\alpha = 0.10$
Step III	State the associated test statistic.	Levene F
Step IV	Identify the random sampling distribution of the test statistic when H_0 is true.	$F \underset{=}{d} F\ (3, 36)$ df when H_0 is true
Step V	State the critical value for rejecting the null hypothesis.	Reject H_0 if $p < .10$
Step VI	Calculate the value of the test statistic from the sample data.	$F = 1.9421$
Step VII	Analyze the results and make an appropriate inference related to the hypotheses tested.	$p = .140$ Discussion:

Figure 11.7 Seven-step Procedure for Hypothesis Testing of Variances.

4, parts produced with coolant supplied from coolant supplier 3 have a statistically significant, higher mean deviation from target than the other suppliers.

Self-Review Activity 11.1 Since we have discussed the basis and theory of statistical inference, you should now be in a position to assess the results of this analysis. Review the statement of the problem, the research hypotheses, the statistical hypotheses, and the results of the statistical analysis. Then, complete the following sequence:

1. Respond to the research hypotheses posed.
2. In general, what would you conclude? What should be done? Why?

SPSS for MS WINDOWS Release 6.1

----- O N E W A Y -----

Variable DEVIATION   Deviation from target
By Variable SUPPLIER

Multiple Range Tests: Student-Newman-Keuls test with significance level .050

The difference between two means is significant if
  MEAN(J)-MEAN(I) >= 1.5222 * RANGE * SQRT(1/N(I) + 1/N(J))
with the following value(s) for RANGE:

Step   2   3   4
RANGE   2.87 3.45 3.81

(*) Indicates significant differences which are shown in the lower triangle

```
                    G G G G
                    r r r r
                    p p p p

                    4 2 1 3
  Mean    SUPPLIER

11.4040   Grp 4
11.5040   Grp 2
12.4393   Grp 1
18.4760   Grp 3      * * *
```

Homogeneous Subsets (highest and lowest means are not significantly different)

Subset 1

Group   Grp 4   Grp 2   Grp 1

Mean   11.4040   11.5040   12.4393
- -

Subset 2

Group   Grp 3

Mean   18.4760

Figure 11.8 Student-Newman-Keuls Procedure (Using SPSS/PC) to Compare the Means of Four Coolant Suppliers.

Case Study 2: The Randomized Block Design: Matched Pairs

The randomized block design is based upon the principle of assigning experimental units to homogeneous groups, or blocks, so that the unit comparison made within blocks is subject to less experimental error than an equivalent comparison across blocks or units (Kirk, 1968). Recall that a blocked factor is not a treatment and the researcher is not interested in increasing the size of the experiment to study the effect. Rather, the researcher's aim is twofold: (1) to study the treatment effect free of the blocked variable, while (2) maintaining the external validity of the comparison by avoiding the option of (block) limitation (Juran, Gryna, and Bingham, 1979). Some authors note that it is also possible that the researcher might wish to obtain an assessment of the influence of the blocked variable, but this observation should be considered with great caution. While estimates of the relative variability effect may be obtained from the statistical analysis, the novice practitioner should be cautioned that the experiment was not structured or designed for the study of the blocked factor. Typically, a randomized block design is employed such that each treatment appears in each block exactly once (Natrella, 1963). In light of these premises, statistical tests of significance should never be calculated or reported for a blocked variable.

Finally, an underlying assumption of a randomized block design is that there is no interaction effect between a treatment effect and a blocked effect. If the researcher has reason to believe that this is not the case, the blocked factor should be incorporated or limited.

The most basic form of a randomized block design is the matched pairs design where the treatment effect consists of only two levels. For our sample problem, we will use the following example:

> A manufacturing firm is considering a new technique and associated device for the monitoring of key and critical equipment in their production facilities. Currently, the maintenance group in each facility monitors bearings and die sets based upon a vibration analysis plan. As vibration of bearings and die sets increases, maintenance employees can determine when to take preventive action by repairing or replacing bearings and die sets before a breakdown occurs. The proposed new monitoring technique uses a patented combination of vibration and heat transfer techniques to perform an equivalent analysis. The process vendor claims that the new device and method is far superior to the existing method in predicting failures. These failures have the effect of increasing maintenance overtime due to catastrophic failures (e.g., bearings seizing).

Of course, the firm's managers recognize that they cannot obtain something for nothing. The new process would require a significant expenditure of funds to install in all of the production facilities and would cost more to operate. On the other hand, the downtime and overtime currently associated with missed predictions are creating internal costs and causing missed promised delivery dates to customers.

The treatment effect in this case is fairly straightforward: monitoring method (*method*). Further, management wants to have the systems compared in all 10 production facilities. With a criterion measure of overtime hours per quarter, and a general recognition that a single process would be selected for all plants, it makes sense to test the two systems in all facilities. On the other hand, there is no wish to expand the study to a size required to test for overtime differences from plant to plant. In fact, we already know such differences exist due to the different mix of equipment in different plants. Given an assumption on the part of the experimental design team that an interaction between plant and method is highly unlikely, we have decided to block out the effects of plant-to-plant variability for this study. All 10 plants, therefore, will still be included in the study without increasing the sample size required to test the treatment.

Statement of the problem: The purpose of the study is to determine whether a significant reduction in overtime hours due to unexpected equipment failures may be achieved by changing the monitoring process from vibration to vibration/heat transfer analysis. The analysis will be conducted across all 10 plants in our system. The order of the treatment level assignment in each plant will be randomly assigned.

Research hypothesis: There is no significant difference in the average overtime hours associated with the use of the current vibration prediction system (V) and the proposed vibration/heat transfer prediction system (VH).

Statistical hypotheses to be tested:

$$H_0: \mu_V = \mu_{VH}$$

$$H_1: \mu_V \neq \mu_{VH}$$

Organization of the independent variables: The management strategies associated with the independent variables corresponding to this study are presented in Fig. 11.9.

Treatment Variable(s)	Method I - Incorporated N - Nested	Classification QN - Quantitative QL - Qualitative	Type F - Fixed R - Random	Number of Levels	Level Description
1. *Prediction Method - Failure Analysis* 2. 3. 4. 5. 6. 7. 8. 9.	*I*	*QL*	*F*	2	Vibration + Heat (1) Vibration only (2; control)
Blocked Variables	**Number and Value for Levels Blocked**	**Variables Limited**		**At Level**	
1. *Plant* 2. 3. 4.	10 Levels - All plant locations	1. None 2. 3. 4.			

Figure 11.9 Organization of Treatment, Block, and Limited Variables for Case Study 2.

The data collected as associated with this study represents hours of overtime during equal test periods and appears as

	Method	
Plant (block)	Vibration/heat (VH)	Vibration (control) (V)
1	450	470
2	320	340
3	580	600
4	570	590
5	600	630
6	380	440
7	470	490
8	510	530
9	420	460
10	380	410

The appropriate analysis of these data may be accomplished with any one of three statistical approaches. The first, and probably most common, would be the matched pairs t test. The first five steps of the seven-step procedure appear in Fig. 11.10.

The matched pairs t test was generated using SPSS as shown in Fig. 11.11. Now, the remaining steps can be added to the seven-step procedure in Fig. 11.12.

The same result may be obtained through ANOVA. In this case, the first five steps of the seven-step procedure appear in Fig. 11.13. The plant effect is separated, even though we are not really testing its effect. This allows us to test the method effect relative to the error term (residual). Results of the statistical analysis (from SPSS/PC) are presented in Fig. 11.14. Note that the paired t test and ANOVA yield equivalent results, since $t^2 = F$. (The software package employed for ANOVA reports the F value and significance of the block effect (plant), but the numbers are meaningless and should not be used in the analysis.)

The seven-step procedure can now be updated (Fig. 11.15) with the results of the analysis. Both tests indicate that we should reject the null hypothesis (that overtime hours due to unexpected equipment failures are the same when monitoring the process using the current vibration method as when using the vibration/heat transfer system).

Self-Review Activity 11.2 To complete this case study:

1. Respond to the research hypothesis posed.
2. In general, what would you conclude? What should be done? Why?
3. Finally, was blocking the plant effect important? If the exact same data had been gathered, with the plant effect ignored and permitted to artificially inflate (confound with) the experimental error term, would we have found a significant difference? Examine

Step Number	Description	Participant Response
Step I	State the null and research hypotheses.	$H_0: \mu_V = \mu_{VH}$ $H_1: \mu_V \neq \mu_{VH}$
Step II	State the maximum risk of committing a Type I error.	$\alpha = 0.10$
Step III	State the associated test statistic.	$t = \dfrac{\overline{D}}{S_D/\sqrt{n}}$
Step IV	Identify the random sampling distribution of the test statistic when H_0 is true.	$t \underline{d} t(9)$ df when H_0 is true
Step V	State the critical value for rejecting the null hypothesis.	Reject H_0 if $p < 0.10$
Step VI	Calculate the value of the test statistic from the sample data.	$t =$
Step VII	Analyze the results and make an appropriate inference related to the hypotheses tested.	$p =$ Discussion:

Figure 11.10 Seven-step Procedure for the Matched Pairs t Test for Means.

SPSS for MS WINDOWS Release 6.1

t-tests for Paired Samples

Variable		Number of pairs	Corr	2-tail Sig	Mean	SD	SE of Mean
		10	.991	.000			
V	Vibration (Control) Method				496.0000	91.676	28.990
VH	Vibration / Heat Transfer Method				468.0000	95.545	30.214

Paired Differences

Mean	SD	SE of Mean	t-value	df	2-tail Sig
28.0000	13.166	4.163	6.73	9	.000

95% CI (18.582, 37.418)

Figure 11.11 Matched Pairs t Test Comparing Average Overtime Hours Due to Unexpected Breakdowns for Two Preventive Maintenance Methods.

Step II	State the maximum risk of committing a Type I error.	$\alpha = 0.10$
Step III	State the associated test statistic.	$t = \dfrac{\overline{D}}{S_D/\sqrt{n}}$
Step IV	Identify the random sampling distribution of the test statistic when H_0 is true.	$t \underset{=}{d} t(9)$ df when H_0 is true
Step V	State the critical value for rejecting the null hypothesis.	Reject H_0 if $p < 0.10$
Step VI	Calculate the value of the test statistic from the sample data.	$t = -6.73$
Step VII	Analyze the results and make an appropriate inference related to the hypotheses tested.	$p = .000$ Discussion: Reject H_0. We have sufficient statistical evidence to infer that there is a difference between the average overtime levels for the two methods.

Figure 11.12 Seven-step Procedure for the Matched Pair t Test of Means.

Step Number	Description	Participant Response
Step I	State the null and research hypotheses.	$H_0: \quad \mu_V = \mu_{VH}$
		$H_1: \quad \mu_V \neq \mu_{VH}$
Step II	State the maximum risk of committing a Type I error.	$\alpha = 0.10$
Step III	State the associated test statistic.	$F = MS_{method} / MS_{res}$
Step IV	Identify the random sampling distribution of the test statistic when H_0 is true.	$F \underset{=}{d} F(1,9)$ df when H_0 is true
Step V	State the critical value for rejecting the null hypothesis.	Reject H_0 if $p < 0.10$
Step VI	Calculate the value of the test statistic from the sample data.	$F =$
Step VII	Analyze the results and make an appropriate inference related to the hypotheses tested.	$p =$ Discussion:

Figure 11.13 Seven-step Procedure for the Mean Comparison Using ANOVA.

SPSS for MS WINDOWS Release 6.1

*** ANALYSIS OF VARIANCE ***

OVERTIME Hours Overtime for Unexpected Breakdowns
by METHOD
    PLANT

UNIQUE sums of squares
All effects entered simultaneously

Source of Variation	Sum of Squares	DF	Mean Square	F	Sig of F
Main Effects	160940.000	10	16094.000	185.700	.000
METHOD	3920.000	1	3920.000	45.231	.000
PLANT	157020.000	9	17446.667	{201.308	.000}*
Explained	160940.000	10	16094.000	185.700	.000
Residual	780.000	9	86.667		
Total	161720.000	19	8511.579		

20 cases were processed.
0 cases (.0 pct) were missing.

* Blocked variable

Figure 11.14 ANOVA Comparing Average Number of Overtime Hours Due to Unexpected Breakdowns for Two Preventive Maintenance Methods.

Chapter Eleven

Step Number	Description	Participant Response
Step I	State the null and research hypotheses.	$H_0: \mu_V = \mu_{VH}$ $H_1: \mu_V \neq \mu_{VH}$
Step II	State the maximum risk of committing a Type I error.	$\alpha = 0.10$
Step III	State the associated test statistic.	$F = MS_{Method} / MS_{Res}$
Step IV	Identify the random sampling distribution of the test statistic when H_0 is true.	$F \underset{=}{d} F(1,9)$ df when H_0 is true
Step V	State the critical value for rejecting the null hypothesis.	Reject H_0 if $p < 0.10$
Step VI	Calculate the value of the test statistic from the sample data.	$F = 45.23$
Step VII	Analyze the results and make an appropriate inference related to the hypotheses tested.	$p = .000$ Discussion: Reject H_0. We have sufficient statistical evidence to infer that there is a difference between the average overtime levels for the two methods.

Figure 11.15 Seven-step Procedure for the Mean Comparison Using ANOVA.

SPSS for MS WINDOWS Release 6.1

t-tests for Independent Samples of METHOD

Variable	Number of Cases	Mean	SD	SE of Mean
OVERTIME Hours Overtime for Unexpected Breakdowns				
Vibration / Heat Transfer	10	468.0000	95.545	30.214
Vibration (Control) Method	10	496.0000	91.676	28.990

Mean Difference = -28.0000

Levene's Test for Equality of Variances: F= .048 P= .829

| t-test for Equality of Means | | | | | 95% |
Variances	t-value	df	2-Tail Sig	SE of Diff	CI for Diff
Equal	-.67	18	.512	41.873	(-115.972, 59.972)
Unequal	-.67	17.97	.512	41.873	(-115.982, 59.982)

Figure 11.16 Independent t Test Comparing Average Number of Overtime Hours Due to Unexpected Breakdowns for Two Preventive Maintenance Methods (without Blocking).

the statistical analysis in Fig. 11.16. It should be compared directly to the matched pairs t test for the data that were previously presented. What are your conclusions? Why was matching necessary? What do you suppose would have happened if five plants had been randomly selected to test the new method and five plants had been selected to test the current method?

Case Study 3: A Randomized Block Design for More Than Two Treatment Levels

A natural extension of the matched pairs design is the classical randomized block design, where the treatment effect is studied at more than two levels. The underlying assumptions and procedures are identical to those detailed in case study 2. This example was suggested by Dowdy and Weardon (1983) in the text *Statistics for Research,* and is reprinted by permission of the publisher, John Wiley and Sons Inc.

Statement of the problem: Four varieties of hybrid corn have been developed for resistance to the fungal infection known as smut. However, nothing is known about their potential for grain yield. Each hybrid is planted at each of five locations within the state, and the following yields (represented by average bushels of corn per acre) are obtained:

Treatment (hybrid)	Block (location)				
	NW	NE	C	SE	SW
FR-11	62.3	64.0	64.3	65.0	66.4
BCM	63.3	62.7	66.2	66.8	64.5
DBC	60.8	64.3	65.2	62.2	65.1
RC-3	55.4	56.0	59.8	58.0	58.8

Research hypothesis: There is no significant difference in average yield between the hybrids tested.

Statistical hypotheses to be tested:

$$H_0: \mu_1 = \mu_2 = \mu_3 = \mu_4$$

$$H_1: \mu_1 \neq \mu_2 \neq \mu_3 \neq \mu_4$$

Organization of the treatment variables: The management strategies associated with the treatment variables corresponding to this study are presented in Fig. 11.17. The statistical analysis corresponding to these data is quite similar to the example previously presented as illustrated by the first five steps of the seven-step procedure in Fig. 11.18. The results of the analysis from SPSS/PC are shown in Fig. 11.19. (Again, software reports F value and significance of block effect but the numbers are meaningless and should not be reported.)

This analysis allows us to present the completed seven-step procedure (Fig. 11.20). The analysis shows $F = 38.97$, which corresponds to $p = .000$. This means we should reject the null hypothesis H_0. We have sufficient statistical evidence to infer that there is a difference in average yield for the four hybrids. The w^2 calculation is a measure of importance. It displays the percent of variation explained by the model (in this case, by the hybrid effect); $w^2 = 74$ per-

	Treatment Variable(s)	Method I - Incorporated N - Nested	Classification QN - Quantitative QL - Qualitative	Type F - Fixed R - Random	Number of Levels	Level Description
1.	Hybrid	I	QL	F	4	Varieties FR-11/BCM/DBC/RC-3
2.						
3.						
4.						
5.						
6.						
7.						
8.						
9.						
	Blocked Variables	**Number and Value for Levels Blocked**		**Variables Limited**		**At Level**
1.	Geographical Location	5 locations: northwest, northeast, central, southeast, southwest		1. None		
2.				2.		
3.				3.		
4.				4.		

Figure 11.17 Organization of Treatment, Block, and Limited Variables for Case Study 3.

cent implies that, not only is the hybrid effect significant, but it has a large (important) effect on yield.

The next obvious question is: If there is a difference between the hybrids, which ones work best? This post-hoc analysis can be addressed as follows. A review of the means shows that hybrids 1, 2, and 3 seem to have yields that are close to equal, while hybrid 4 has a lower sample average yield. The statistical analysis presented in Fig. 11.21 shows (through a one-way ANOVA of hybrids 1, 2, and 3) that there is no significant difference between average yield for hybrids 1, 2, and 3. Therefore, the difference identified in Fig. 11.19 must be due to hybrid 4 having a significantly lower yield than hybrids 1, 2, and 3.

Self-Review Activity 11.3 Given the statistical analysis and results presented in this case study:

1. Respond to the research hypothesis posed.
2. In general, what would you conclude? Are there important differences in yield among the means of the hybrids? Why?
3. Which of the hybrids might you recommend? Why?

Step Number	Description	Participant Response
Step I	State the null and research hypotheses.	$H_0:$ $\mu_1 = \mu_2 = \mu_3 = \mu_4$
		$H_1:$ $\mu_1 \neq \mu_2 \neq \mu_3 \neq \mu_4$
Step II	State the maximum risk of committing a Type I error.	$\alpha = 0.10$
Step III	State the associated test statistic.	$F = MS_{Hybrid} / MS_{Res}$
Step IV	Identify the random sampling distribution of the test statistic when H_0 is true.	$F \stackrel{d}{=} F(3, 12)$ df when H_0 is true
Step V	State the critical value for rejecting the null hypothesis.	Reject H_0 if $p < 0.10$
Step VI	Calculate the value of the test statistic from the sample data.	$F =$
Step VII	Analyze the results and make an appropriate inference related to the hypotheses tested.	$p =$ Discussion:

Figure 11.18 Seven-step Procedure for Mean Comparison for the Effect of Four Hybrids on Yield.

SPSS for MS WINDOWS Release 6.1

Group	Count	Mean	Standard Deviation	Standard Error	95 Pct Conf Int for Mean
Grp 1	5	64.4000	1.4950	.6686	62.5438 TO 66.2562
Grp 2	5	64.7000	1.7790	.7956	62.4911 TO 66.9089
Grp 3	5	63.5200	1.9409	.8680	61.1101 TO 65.9299
Grp 4	5	57.6000	1.8601	.8319	55.2904 TO 59.9096
Total	20	62.5550	3.3869	.7573	60.9699 TO 64.1401

*** ANALYSIS OF VARIANCE ***

YIELD
by HYBRID
LOCATION

UNIQUE sums of squares
All effects entered simultaneously

Source of Variation	Sum of Squares	DF	Mean Square	F	Sig of F	
HYBRID	167.442	3	55.814	38.972	.000	$w^2 = 74.37\%$
LOCATION (BLOCK)	33.322	4	8.331	{5.817	.008}	
Residual	17.186	12	1.432			

20 cases were processed.
0 cases (.0 pct) were missing.

Figure 11.19 Descriptive Statistics and ANOVA to Compare Mean Yield for Four Hybrids (Blocking location).

Chapter Eleven

Step Number	Description	Participant Response
Step I	State the null and research hypotheses.	$H_0: \mu_1 = \mu_2 = \mu_3 = \mu_4$ $H_1: \mu_1 \neq \mu_2 \neq \mu_3 \neq \mu_4$
Step II	State the maximum risk of committing a Type I error.	$\alpha = 0.10$
Step III	State the associated test statistic.	$F = {MS_{Hybrid}} / {MS_{Res}}$
Step IV	Identify the random sampling distribution of the test statistic when H_0 is true.	$F \underset{=}{d} F(3, 12)$ df when H_0 is true
Step V	State the critical value for rejecting the null hypothesis.	Reject H_0 if $p < 0.10$
Step VI	Calculate the value of the test statistic from the sample data.	$F = 38.97$
Step VII	Analyze the results and make an appropriate inference related to the hypotheses tested.	$p = .000$ Discussion: Reject H_0. We have sufficient statistical evidence to infer that there is a difference in average yield for the four hybrids.

Figure 11.20 Seven-step Procedure for Mean Comparison for the Effect of Four Hybrids on Yield.

SPSS for MS WINDOWS Release 6.1

*** ANALYSIS OF VARIANCE ***

YIELD
by HYBRID
LOCATION

UNIQUE sums of squares
All effects entered simultaneously

Source of Variation	Sum of Squares	DF	Mean Square	F	Sig of F
HYBRID	3.761	2	1.881	.984	.415
LOCATION	21.376	4	5.344	2.796	.101
Residual	15.292	8	1.911		

20 cases were processed.
0 cases (.0 pct) were missing.

Figure 11.21 Post-hoc Analysis—ANOVA to Compare Mean Yield for Hybrids 1, 2, and 3. (Blocking location)

Case Study 4: A Complete Block Design for More Than Two Blocked Effects

Latin square design

In some cases, we will wish to conduct an experiment for a single treatment, where we simultaneously have two blocked effects or factors. One of the designs that may be used to accomplish this task is the Latin square. Much of the early work in the design and utilization of the Latin square was conducted in agriculture. However, many groups in the automotive industry have employed this design procedure for a number of years. A classical example of this design, from work conducted in the early 1970s at Ford Motor Company, will be used to illustrate the nature and purpose of this design.

We begin with a desire on the part of an automotive company to test the resistance to tire wear on cars built by the company. Four tire manufacturers are to be considered. Let the terms A, B, C, and D represent the four levels of the treatment (the four tire manufacturers). The firm has four car models on which the tires might be tested.

By now, of course, you recognize that the layout in Table 11.1 would never be employed for this study. This design would be totally unacceptable as a means of answering this research question. We can observe from the layout that our

TABLE 11.1 Initial Design Plan (Flawed).

Car Model		1	2	3	4
Tire Manufacturer		A	B	C	D
Tire Wear Value (OBS)	1	X_{11}	X_{12}	X_{13}	X_{14}
	2	X_{21}	X_{22}	X_{23}	X_{24}
	3	X_{31}	X_{32}	X_{33}	X_{34}
	4	X_{41}	X_{42}	X_{43}	X_{44}

treatment, tire manufacturer, would be hopelessly confounded with the car used to test the four sets of tires. There would be no way, in such a case, to determine whether any observed differences were due to the differences in tires, or in the cars.

This condition could be avoided, of course, by incorporating the car effect. Management, however, has indicated that they already know that different cars yield different degrees of wear and they do not wish to pay for another experiment to document that phenomenon. Further, during discussions related to planning for this experiment, they have warned against limiting our study to a single car model. They wish to select one tire manufacturer for the models under consideration, and they want the vehicle variability as represented by the four car models included in the experiment. Obviously, if we must include each of the four types of vehicles in the experiment, do not wish to confound the levels of car with the treatment levels, and do not wish to incorporate car into our study, the solution would be to block the effects of the four car models. This could be accomplished by assigning one of each manufacturer's tires to each of the four vehicles (Table 11.2).

This design looked fairly effective, until the experimental design team continued to plan the experiment. It then occurred to the team members that the tire position could be a confounding factor. Suppose tire wear varies by tire position (it does). Suppose further that, for example, manufacturer A's tires happened to end up on the right front and perhaps low wear position for three or four of the vehicles. The resultant low level of tire wear would be incorrect-

TABLE 11.2 Secondary Design Plan.

	Tire Manufacturer	A	B	C	D
	1	X_{11}	X_{12}	X_{13}	X_{14}
Car	2	X_{21}	X_{22}	X_{23}	X_{24}
	3	X_{31}	X_{32}	X_{33}	X_{34}
	4	X_{41}	X_{42}	X_{43}	X_{44}

TABLE 11.3 Final Design Plan: A Latin Square.

Tire Position		LF	LR	RF	RR
	1	A	D	C	B
Car	2	B	A	D	C
	3	C	B	A	D
	4	D	C	B	A

ly attributed to manufacturer A's "superior" tires. The design team returned to the management group, and was told that (1) we already know that tires wear unevenly by position, (2) we do not want the experiment to test this effect again, and (3) make certain that the design has the external validity necessary to select the nominal (lowest wear) tire for all four wheel positions, and of course, for all vehicles in the model class tested.

The design team returned to their work area and selected a Latin square design plan. Table 11.3 illustrates the value of the Latin square. It allows us to test our single treatment effect, free of confounded extraneous variable effects, yet, without the validity threats of limitation or the additional expense of incorporation or nesting.

Examine the design layout carefully. We may note that the following conditions and assumptions apply as related to this type of design:

1. Each treatment appears in each column and row only once.
2. The number of levels for the treatment and block effects must be the same. In other words, the number of treatment levels must equal the number of columns and must equal the number of rows.
3. It is assumed that there is no interaction present between the three effects or that the interaction, if present, is incorporated statistically and/or technically into the research questions. Whatever treatment effect is present, we assume that it is uniform across all rows and columns.
4. Note that we can test four treatment levels, across four cars, and four position levels, without increasing the sample size necessary to conduct the study of the treatment. This makes the Latin square an extremely useful and efficient design.
5. Finally, you should be aware that Latin square designs are represented in a number of sizes. Figure 11.22 presents a number of Latin squares of various sizes (Natrella, 1963).

The analysis of the Latin square design proceeds in the same fashion as the randomized block design except, of course, that there are two blocked factors instead of one. Let us examine the use of this design by reviewing the following background statement from a sales and marketing experiment.

Background of the problem: A marketing expert for a food products company wishes to measure consumer preference for five different label designs on a particular package. The product is manufactured in New York City and predomi-

3 x 3
```
A B C
B C A
C A B
```

4 x 4
```
A B C D
B D A C
C A D B
D C B A
```

5 x 5
```
A B C D E
B A E C D
C D A E B
D E B A C
E C D B A
```

6 x 6
```
A B C D E F
B F D C A E
C D E F B A
D A F E C B
E C A B F D
F E B A D C
```

7 x 7
```
A B C D E F G
B C D E F G A
C D E F G A B
D E F G A B C
E F G A B C D
F G A B C D E
G A B C D E F
```

8 x 8
```
A B C D E F G H
B C D E F G H A
C D E F G H A B
D E F G H A B C
E F G H A B C D
F G H A B C D E
G H A B C D E F
H A B C D E F G
```

Figure 11.22 Selected Latin Square Designs.

nantly sold in five major outlets. The criterion measure for consumer preference is measured by sales level for the package in the first week of exposure.

The problem the researcher faces is that she wants to test five different label designs, across all of the company's five outlets, for the entire city to determine which label is best overall (no matter what borough or outlet). However, she knows from past research that significantly different sales levels might be expected from borough to borough. She wants to test the five label designs "free" of the borough effects. She also knows that only one design will be selected for all outlets and that the different outlets represent significantly different sales levels from week to week (regardless of and uniformly across the boroughs).

She selects a 5 × 5 Latin square design to conduct her experiment. Figure 11.23 presents her organization of the independent variables.

Designing the Industrial Experiment

Figure 11.23 Organization of Treatment, Block, and Limited Variables for Case Study 4.

Treatment Variable(s)	Method I - Incorporated N - Nested	Classification QN - Quantitative QL - Qualitative	Type F - Fixed R - Random	Number of Levels	Level Description
1. Label Design *Design Legend:* BR - Bright red LR - Light red BB - Bright blue LB - Light blue	I	QL	F	5	BR/BB/Matte BR/BB/Glossy LR/LB/Matte LR/LB/Glossy BR/LB/Matte

Blocked Variables	Number and Value for Levels Blocked	Variables Limited	At Level
1. Borough of NYC	Bronx Brooklyn Queens Manhattan Richmond (S.I.)	1.	
2. Outlet	A-Store Buy Here Cost Saver Deliver-to-You EZ Shop	2.	
3.		3.	
4.		4.	

TABLE 11.4 Data Collected for Label Analysis—Case Study 4.

	New York City Borough				
Outlet	Bronx	Brooklyn	Queens	Manhattan	Richmond
A-Store	205 (D)	295 (C)	275 (A)	225 (E)	270 (B)
Buy Here	265 (C)	275 (B)	240 (E)	270 (A)	250 (D)
Cost Saver	250 (A)	235 (D)	275 (C)	260 (B)	175 (E)
Deliver-to-You	255 (B)	195 (E)	235 (D)	210 (C)	205 (A)
EZ Shop	330 (E)	330 (A)	275 (B)	275 (D)	225 (C)

Treatment Levels: A - BR/BB/M; B - BR/BB/G; C - LR/LB/M; D - LR/LB/G; E - BR/LB/M

Table 11.4 presents the layout of the Latin square employed, and the data collected in conjunction with the study.

Self-Review Activity 11.4 Given that the analysis of this design is consistent with the randomized block design, you should be prepared to complete some of the details related to this study on your own. Complete the following sections:

1. Define the statement of the problem.
2. Define the research hypothesis(es).
3. List the statistical hypothesis(es) to be tested.
4. Complete the first five steps of the seven-step procedure. (Blank forms appear in App. D.)

SPSS for MS WINDOWS Release 6.1

Summaries of   SALES $ Sales in First Week
By levels of   LABEL   Label Design

Variable	Value	Label	Mean	Std Dev	Cases
For Entire Population			252.0000	38.2426	25
LABEL	1.00	BR/BB/M	266.0000	45.1940	5
LABEL	2.00	BR/BB/G	267.0000	9.0830	5
LABEL	3.00	LR/LB/M	254.0000	35.4260	5
LABEL	4.00	LR/LB/G	240.0000	25.4951	5
LABEL	5.00	BR/LB/M	233.0000	59.8540	5

Total Cases = 25

*** ANALYSIS OF VARIANCE ***

by
SALES $ Sales in First Week
LABEL   Label Design
LOCATION New York City Borough
OUTLET   Sales Outlet

UNIQUE sums of squares
All effects entered simultaneously

Source of Variation	Sum of Squares	DF	Mean Square	F	Sig of F
Main Effects	22510.000	12	1875.833	1.788	.164
LABEL	4650.000	4	1162.500	1.108	.397
LOCATION	5430.000	4	1357.500	{1.294	.327}
OUTLET	12430.000	4	3107.500	{2.962	.065}
Explained	22510.000	12	1875.833	1.788	.164
Residual	12590.000	12	1049.167		
Total	35100.000	24	1462.500		

25 cases were processed.
0 cases (.0 pct) were missing.

Figure 11.24 Statistical Analysis (Means and ANOVA) for Identifying Customer Preference for Label Design.

Given that you have provided the prose, the statistical analysis for this study appears in Fig. 11.24.

Self-Review Activity 11.5

1. Values bracketed { } are provided automatically by the software. How should these be interpreted?
2. What can we conclude from this study? Did you select a label design? Why or why not? If you did select a particular design, which one?

Youden square designs and other variations of the Latin square design

Before progressing to experimental designs involving more than one factor, two final notes should be reviewed regarding the Latin square and associated extensions of these plans. One of the restrictions of a Latin square is that the number of levels associated with the treatment and block effects must be equal. In cases where the number of treatment levels is equal to the number of levels of one of the block effects, but greater than the number of levels for the second block effect, a Youden Square may be employed. These designs are structured and interpreted exactly the same as Latin squares, with the exception of the lowered restriction. For example, a Plan 5 Youden Square allows for a treatment and block effect to be tested at seven levels each, with a second block effect tested at three levels. For a comprehensive description of the construction of Youden square plans, refer to *Quality Control Handbook* (Juran, Gryna, and Bingham, 1979).

Finally, Graeco-Latin square and hyper-Graeco-Latin square plans also exist. They are employed for the study of a single factor with three or more than three (respectively) blocked effects. Box, Hunter, and Hunter (1978) provide a complete discussion of these designs.

Case Study 5: Factorial Experiments—A Fully Crossed Type I Model

Factorial experiment is a categorical term assigned to those experiments where more than one treatment effect is studied at two or more levels. The experimental plan generally consists of testing n experimental units at each treatment combination, for all combinations of the treatment levels. For this reason, factorial designs are also often referred to as *full factorial* studies, thereby distinguishing them from *fractional factorial* experiments.

These experiments, and their associated analyses, can become extremely complex and require discussions far beyond the scope of this book. Three basic examples of complete, or full, factorial designs will be discussed in this chapter.

The simplest factorial design layout that may be utilized for a discussion of factorial analyses is the 2^2 factorial design. The nomenclature "2^2" is a com-

TABLE 11.5 Sample Experimental Design Layout for Two Treatment Variables (Crossed Design).

		Treatment Variable 'A' Vendor		
		Acme	Southco	Newmark
Treatment Variable 'B'	#1	10	15	20
Production Line	#2	17	22	27

mon method of describing a factorial study. The first number, 2 in this case, indicates the number of levels associated with the superscript, which represents the number of treatment factors in the study. A 3^2 design, therefore, would indicate a study consisting of 2 treatments, each to be studied at 3 levels. A 2^3 factorial design would indicate a design containing 3 treatment factors, each of which is to be studied at 2 levels.

When we begin discussing effects to be analyzed in a completely randomized factorial design, we also refer to testing main effects and interaction effects. The *main effect* for a specific treatment variable refers to the difference in the levels for that incorporated factor across all other effects. *Interaction* refers to the condition where the difference in the response between levels for a given factor varies according to the value of a second factor.

Consider the factorial design for two treatment variables that we reviewed in Chap. 5 (presented in Table 11.5).

In this case, the matrix has been completed with cell means (sample averages). As you review this figure, note the statistical hypotheses to be tested for this design.

Statistical hypotheses to be tested:

1. H_0: $\mu_{ACME} = \mu_{SOUTHCO} = \mu_{NEWMARK}$

   H_1: $\mu_{ACME} \neq \mu_{SOUTHCO} \neq \mu_{NEWMARK}$

2. H_0: $\mu_{LINE1} = \mu_{LINE2}$

   H_1: $\mu_{LINE1} \neq \mu_{LINE2}$

3. H_0: $I_{AB} = 0$

   H_1: $I_{AB} \neq 0$

where each cell contains n sample observations.

If we were to plot the cell means, we would find that there was no interaction effect between the two main effects. Refer to Fig. 11.25 for a display of these data. Note, for example, that although there is a difference (assume that statistical tests have been conducted and reflect deviations greater than that

Sample Cell Averages

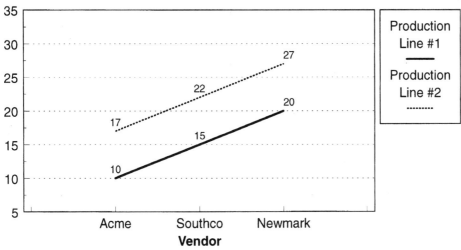

Figure 11.25 Plot of Cell Means for Factorial Design.

TABLE 11.6 Sample Experimental Design Layout for Two Treatment Variables (Crossed Design).

		Treatment Variable 'A' Vendor		
		Acme	Southco	Newmark
Treatment Variable 'B'	#1	10	15	27
Production Line	#2	17	22	20

which would be expected due to experimental error) between the two production lines, the difference is uniform regardless of the level of vendor studied.

Now, review the modified data set created and displayed in Table 11.6. These data have been used to generate the plot displayed in Fig. 11.26.

Note the difference between this plot and the previous display. Consider also the problem, as related to Fig. 11.26, of testing a hypothesis that questions whether (for this example) "production line 1 is different than production line 2, regardless of vendor." What are your conclusions regarding the analysis of main effects in the presence of significant interactions?

Incidentally, you should note that these visual examples are presented to illustrate the concept of the nature of an interaction effect, not the method by which we test for an interaction. Since the plotted points are sample means, they are subject to experimental error. What appears as a set of nonparallel lines when plotted, therefore, could be due to sampling error alone. It is the

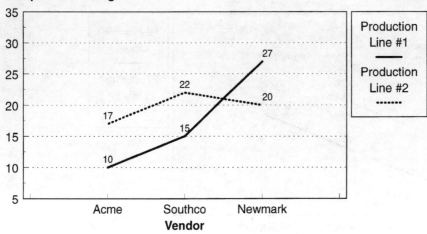

Figure 11.26 Plot of Cell Means for Factorial Design.

statistical analysis associated with the ANOVA that will tell us that the observed interaction is likely not due to experimental error alone, not a simple plot.

In order to illustrate the use and analysis of this type of design, we will review an experiment designed to increase the adherence properties of magnets glued into metal "cups" as part of the manufacturing process for electric motors which regulate windows in automobiles. In this study, two treatment variables will be studied at three levels each. Figure 11.27 reflects the organization of the experimental design.

Treatment Variable(s)	Method I - Incorporated N - Nested	Classification QN - Quantitative QL - Qualitative	Type F - Fixed R - Random	Number of Levels	Level Description
1. *Glue Type*	*I*	*QL*	*F*	3	*1 - Current* *2 - TRT A* *3 - TRT B*
2. *Glue Temperature*	*I*	*QN*	*F*	3	*1 - Low* *2 - Target* *3 - High*
3.					
4.					
5.					
Blocked Variables	**Number and Value for Levels Blocked**		**Variables Limited**		**At Level**
1. None 2. 3. 4.			1. None 2. 3. 4.		

Figure 11.27 Organization of Treatment, Block, and Limited Variables for Case Study 5.

Statement of the problem: The purpose of this study is to determine whether either (or both) of two proposed types of glue is superior to the glue currently used in adherence strength. Additionally, this study is intended to determine whether a significant difference exists in the adherence strength based upon the temperature employed during the gluing process.

Finally, if significant differences or interactions are found to exist, the purpose of this study is to attempt to identify a potential optimum target value for one or both variables in order to maximize adhesion. Adhesion will be measured on the basis of the force required to break the glue seal between the magnet and cup, as measured on a force gauge.

Research hypotheses: There is no significant difference in adhesion strength for magnets glued with our current glue (GC) versus the two new proposed glues (GX1 and GX2).

Adhesion strength of magnets glued at the current target glue temperature (TT) is higher than at the low (TL) and high (TH) extreme levels of the temperature specification.

Statistical hypotheses to be tested:

1. H_0: $\mu_{GC} = \mu_{GX1} = \mu_{GX2}$

    H_1: $\mu_{GC} \neq \mu_{GX1} \neq \mu_{GX2}$

2. H_0: $\mu_{TL} = \mu_{TT} = \mu_{TH}$

    H_1: $\mu_{TL} \neq \mu_{TT} \neq \mu_{TH}$

3. H_0: $I_{GT} = 0$

    H_1: $I_{GT} \neq 0$

Where I_{GT} represents the interaction between glue type (G) and temperature set point (T).

The analysis of the data collected appears in Fig. 11.28.

The ANOVA table in Fig. 11.28 shows that, since the F value for the interaction effect is 1.645 and corresponds to a p value of .160, the interaction effect is not significant. This means that each main effect can be assessed across the levels of the other effect. For glue type, the F value is 321.35 and corresponds to a p value of .000 and for temperature, the F value is 22.671 and corresponds to a p value of .000. In both cases, this indicates that we have sufficient statistical evidence to infer that glue type and temperature both make a difference in adherence strength. The next question is: Even though the main effects (glue and temperature) are significant, are they important? This can be answered by the multiple classification analysis and calculation of η values presented in Fig. 11.29. η^2 is the percent of variation explained by the model.

The multiple classification analysis indicates an η value of .15 and therefore $\eta^2 = .02$ (2 percent) for temperature, and an η value of .57 and therefore, $\eta^2 =$

226 Chapter Eleven

GLTEMP   GLUE - IV1

Value Label	Value	Frequency	Percent	Valid Percent	Cum Percent
LOWER SPECIFICATION	1.00	432	33.3	33.3	33.3
TARGET VALUE	2.00	432	33.3	33.3	66.7
UPPER SPECIFICATION	3.00	432	33.3	33.3	100.0
Total		1296		100.0	100.0

Valid cases   1296   Missing cases 0

GLUE   GLUE TYPE - IV2

Value Label	Value	Frequency	Valid Percent	Cum Percent
CURRENT GLUE	1.00	432	33.3	33.3
PROPOSED TYPE A	2.00	432	33.3	66.7
PROPOSED TYPE B	3.00	432	33.3	100.0
Total		1296	100.0	100.0

Valid cases   1296   Missing cases 0

CELL MEANS

```
    FORCE              FORCE
   BY GLTEMP       GLUE TEMPERATURE
     GLUE             GLUE TYPE
```

TOTAL POPULATION

  .88
(1296)

GLTEMP
   1   2   3

  .97   .86   .80
(432) (432) (432)

Figure 11.28 Full Factorial ANOVA to Determine Effect of Glue Type, Temperature, and Interaction on Adherence Strength.

```
          1      2      3
        1.22    .62    .79
       (432)  (432)  (432)


        GLUE
          1      2      3
GLTEMP
   1    1.29    .73    .88
       (144)  (144)  (144)

   2    1.19    .63    .75
       (144)  (144)  (144)

   3    1.19    .50    .73
       (144)  (144)  (144)
```

ANALYSIS OF VARIANCE

```
         FORCE     FORCE
   BY    GLTEMP    GLUE TEMPERATURE
         GLUE      GLUE TYPE
```

Source of Variation	Sum of Squares	DF	Mean Square	F	Signif of F
Main Effects					
GLTEMP	5.909	2	2.954	22.671	.000
GLUE	83.754	2	41.877	321.350	.000
2-way Interactions					
GLTEMP GLUE	.858	4	.214	1.645	.160
Residual	167.717	1287	.130		
Total	258.238	1295	.199		

Figure 11.28 (*Continued*)

.32 (32 percent) for glue type. This means that glue type is both significant and important. And, although temperature is significant, it is not important—it does not have nearly as large an effect as glue type. Now, since each factor occurs at three levels, the next step would be a post-hoc analysis to determine which levels of glue type give the highest adherence strength.

Before we discuss the process by which we utilize the statistical analysis, we should designate the model with which we are working.

- The model represents a fully crossed design.
- The statistical analysis is referred to as a Model I ANOVA.

In this type of analysis, we generally employ the following procedure in testing the statistical hypotheses:

*** MULTIPLE CLASSIFICATION ANALYSIS ***

```
        FORCE      FORCE
By      GLTEMP     GLUE TEMPERATURE
        GLUE       GLUE TYPE
```

Grand Mean = .876

Variable + Category	N	Unadjusted Dev'n	Eta
GLTEMP			
1 LOW	432	.09	
2 TARGET	432	-.02	
3 HIGH	432	-.07	
			.15
GLUE			
1 CURRENT	432	.35	
2 PROPOSED 'A'	432	-.26	
3 PROPOSED 'B'	432	-.09	
			.57

Multiple R Squared   .347
Multiple R   .589

Figure 11.29 Multiple Classification Analysis for the Effect of Glue Type, Temperature, and Interaction on Adherence Strength.

1. Test the hypothesis associated with the interaction effect.
2. If the hypothesis of no interaction is accepted, review the main effects F tests. If these are significant, and there are more than two levels associated with the significant effect, we proceed to an appropriate post-hoc analysis such as the Student-Neumann-Keuls test or Scheffe's procedure.
3. If the hypothesis of no interaction is rejected, and if both treatments are fixed effects, then the main effects cannot be tested across one another. We would typically plot the cell means associated with the design matrix, and design a comprehensive alternative analytical plan and approach.

The scope of this text prohibits us from delving very far into the intricacies of post-hoc analyses or within-level analyses in the presence of interaction effects. Rather, we will concern ourselves with a discussion of the initial or overall results.

Self-Review Activity 11.6 Returning to our example, we can test the statistical and research hypotheses previously posed:

1. Complete a seven-step procedure for each of the statistical hypotheses posed. (Blank seven-step procedure forms are in App. D.)

2. Respond to the research hypotheses, and provide your conclusions in light of the original statement of the problem.

Case Study 6 and Self-Review Activity 11.7: Factorial Experiments—A Fully Crossed Type I Model

Now that we have reviewed an example of a Model I, fully crossed, completely randomized factorial design, you should be able to analyze the following yourself.

A manufacturing firm in Lincoln, Nebraska, produces injected molded parts for the automotive industry. The firm would like to make one of the critical characteristics associated with this product, the density of the product, as high as possible without leaving the operating envelope of the current process.

A recent screening experiment has indicated that the two critical first-order process variables associated with this criterion measure may be (1) the percentage of filler mixed with the raw polymer and (2) the cycle time associated with the production rate of the process. A complete factorial design has been conducted to test this assumption and to find the optimum set points or targets (if the two variables are shown to be significant and important). The organization of the independent variables associated with this study is detailed in Fig. 11.30.

Given the background, complete the following:

1. Identify the statement of the problem.
2. Define the research hypotheses.
3. List the statistical hypotheses to be tested.

The statistical analysis associated with this design is shown in Fig. 11.31. The analysis is divided into two sections. "Section 1—Means Analysis" shows the means by cycle, by filler, and by cycle and filler combination, the two-way

Treatment Variable(s)	Method I - Incorporated N - Nested	Classification QN - Quantitative QL - Qualitative	Type F - Fixed R - Random	Number of Levels	Level Description
1. *Filler*	*I*	*QN*	*F*	2	*8% / 12%*
2. *Cycle Time*	*I*	*QN*	*F*	4	*6 / 8 / 10 / 12*
3.					
4.					
5.					
6.					
7.					
8.					
Blocked Variables	**Number and Value for Levels Blocked**		**Variables Limited**		**At Level**
1. *None*			1. *Plant*		*Lincoln*
2.			2. *Polymer*		*Low density polyethylene*
3.			3.		
4.			4.		

Figure 11.30 Organization of Treatment, Block, and Limited Variables for Case Study 6.

Means Analysis:

Summaries of DENSITY   PART DENSITY (CRITERION X)
By levels of   FILLER   PERCENTAGE OF FILLER
        CYCLE   CYCLE TIME

Variable	Value	Label	Mean	Variance	Cases
For Entire Population			21.0156	9.1902	64
FILLER	1.00	8%	21.5313	8.9667	32
CYCLE	1.00	6	25.8750	.4107	8
CYCLE	2.00	8	22.3750	.8393	8
CYCLE	3.00	10	19.2500	.2143	8
CYCLE	4.00	12	18.6250	.2679	8
FILLER	2.00	12%	20.5000	9.1613	32
CYCLE	1.00	6	16.1250	.4107	8
CYCLE	2.00	8	24.1250	.9821	8
CYCLE	3.00	10	21.2500	.7857	8
CYCLE	4.00	12	20.5000	.8571	8

Total Cases = 64

CELLMEANS

       DENSITY   PART DENSITY (CRITERION X)
BY   FILLER   PERCENTAGE OF FILLER
      CYCLE   CYCLE TIME

TOTAL POPULATION

    21.02
    (64)

FILLER
    1   2

   21.53   20.50
   (32)   (32)

Figure 11.31 Statistical Analysis for Case Study 6. The Effect of Filler and Cycle Time on Part Density.

| | 21.00 | 23.25 | 20.25 | 19.56 |
| | (16) | (16) | (16) | (16) |

	CYCLE			
	1	2	3	4
FILLER				
1	25.88	22.38	19.25	18.63
	(8)	(8)	(8)	(8)
2	16.13	24.13	21.25	20.50
	(8)	(8)	(8)	(8)

ANALYSIS OF VARIANCE

	DENSITY	PART DENSITY (CRITERION X)
BY	FILLER	PERCENTAGE OF FILLER
	CYCLE	CYCLE TIME

Source of Variation	Sum of Squares	DF	Mean Square	F	Signif of F
Main Effects					
FILLER	17.016	1	17.016	28.551	.000
CYCLE	123.047	3	41.016	68.820	.000
2-way Interactions					
FILLER CYCLE	405.547	3	135.182	226.823	.000
Residual	33.375	56	.596		
Total	578.984	63	9.190		

64 Cases were processed.
0 Cases (.0 PCT) were missing.

MULTIPLE CLASSIFICATION ANALYSIS

	DENSITY	PART DENSITY (CRITERION X)
By	FILLER	PERCENTAGE OF FILLER
	CYCLE	CYCLE TIME

Grand Mean = 21.016

Variable + Category	N	Unadjusted Dev'n	Eta
FILLER			
1 8%	32	.52	
2 12%	32	-.52	

Figure 11.31 (*Continued*)

```
CYCLE
1       6        16      -.02
2       8        16      2.23
3       10       16      -.77
4       12       16      -1.45
                                .46

Multiple R Squared          .242
Multiple R                  .492
```

Variance Analysis:

CELLMEANS

```
    LOGVAR          LOG VARIANCE DENSITY
    BY FILLER       PERCENTAGE OF FILLER
       CYCLE        CYCLE TIME
```

TOTAL POPULATION

   -.67
  (16)

FILLER
    1   2

   -.96   -.38
   (8)   (8)

CYCLE
    1   2   3   4

  -1.07 -.15   -.79   -.66
   (4) (4)   (4)   (4)


```
 CYCLE
         1     2     3     4
FILLER
  1    -.90  -.32 -1.39 -1.24
       ( 2)  ( 2) ( 2)  ( 2)

  2   -1.24   .03  -.20  -.09
       ( 2)  ( 2) ( 2)  ( 2)
```

Figure 11.31 (*Continued*)

ANALYSIS OF VARIANCE

```
    LOGVAR        LOG VARIANCE DENSITY
    BY FILLER     PERCENTAGE OF FILLER
       CYCLE      CYCLE TIME
```

Source of Variation	Sum of Squares	DF	Mean Square	F	Signif of F
Main Effects					
FILLER	1.368	1	1.368	4.898	.058
CYCLE	1.797	3	.599	2.145	.173
2-way Interactions					
FILLER CYCLE	1.608	3	.536	1.920	.205
Explained	4.773	7	.682	2.442	.117
Residual	2.234	8	.279		
Total	7.007	15	.467		

MULTIPLE CLASSIFICATION ANALYSIS

```
    LOGVAR        LOG VARIANCE DENSITY
    By FILLER     PERCENTAGE OF FILLER
       CYCLE      CYCLE TIME
```

Grand Mean = −.669

Variable + Category	N	Unadjusted Dev'n	Eta
FILLER			
1 8%	8	−.29	
2 12%	8	.29	
			.44
CYCLE			
1 6	4	−.40	
2 8	4	.52	
3 10	4	−.13	
4 12	4	.00	
			.51
Multiple R Squared			.452
Multiple R			.672

Figure 11.31 (*Continued*)

Mean Density Values

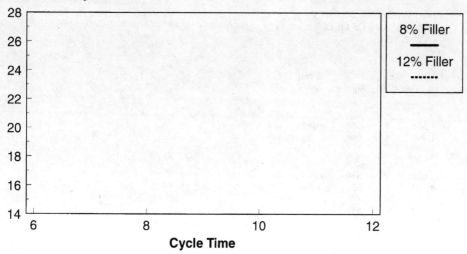

Figure 11.32 Interaction Analysis for Plastic Density—2 × 4 Fully Crossed Factorial Design, Model I.

ANOVA table, and the multiple classification analysis for part density. "Section 2—Variance Analysis" shows the same information for the natural log of the variance ($\ln \sigma^2$) which is the appropriate criterion measure for testing the effects of cycle and filler on the variability of part density.

4. Use the blank graph provided (see Fig. 11.32) to plot the cell means.
5. Completed seven-step procedure forms are displayed in Fig. 11.33. Based on the statistical analysis and the seven-step procedure forms, what are your conclusions? Do you believe that we will be able to find an appropriate and cost-effective combination of set points?

Step Number	Description	Participant Response
Step I	State the null and research hypotheses.	H_0 : $I_{FC} = 0$ H_1 : $I_{FC} \neq 0$
Step II	State the maximum risk of committing a Type I error.	$\alpha = 0.10$
Step III	State the associated test statistic.	$F = MS_F / MS_{Res}$
Step IV	Identify the random sampling distribution of the test statistic when H_0 is true.	$F \underset{=}{d} F(3, 56)$ df when H_0 is true
Step V	State the critical value for rejecting the null hypothesis.	Reject H_0 if $p < 0.10$
Step VI	Calculate the value of the test statistic from the sample data.	$F = 226.823$
Step VII	Analyze the results and make an appropriate inference related to the hypotheses tested.	$p = .000$ Discussion: Reject H_0. We have sufficient statistical evidence to infer that there is a significant interaction between percent filler and cycle time.

Figure 11.33 Seven-step Procedures for Hypothesis Testing for Self-Review Activity 11.7.

Step Number	Description	Participant Response
Step I	State the null and research hypotheses.	$H_0: \mu_{F1} = \mu_{F2}$ $H_1: \mu_{F1} \neq \mu_{F2}$
Step II	State the maximum risk of committing a Type I error.	$\alpha = 0.10$
Step III	State the associated test statistic.	$F = MS_F / MS_{Res}$
Step IV	Identify the random sampling distribution of the test statistic when H_0 is true.	$F \stackrel{d}{=} F(1, 56)$ df when H_0 is true
Step V	State the critical value for rejecting the null hypothesis.	Reject H_0 if $p < 0.10$
Step VI	Calculate the value of the test statistic from the sample data.	$F = 28.551$
Step VII	Analyze the results and make an appropriate inference related to the hypotheses tested.	$p = .000$ Discussion: We cannot assess the filler effect across the cycle effect since there is a significant interaction. The appropriate hypotheses (based on the graph) of means are: 1) $H_0: \mu_{F1} = \mu_{F2}$ $H_0: \mu_{F1} \neq \mu_{F2}$ within cycle time 6 and 2) $H_0: \mu_{F1} = \mu_{F2}$ $H_0: \mu_{F1} \neq \mu_{F2}$ within cycle times 8, 10, 12

Figure 11.33 (*Continued*)

Step Number	Description	Participant Response
Step I	State the null and research hypotheses.	$H_0: \mu_{c1} = \mu_{c2} = \mu_{c3} = \mu_{c4}$ $H_1: \mu_{c1} \neq \mu_{c2} \neq \mu_{c3} \neq \mu_{c4}$
Step II	State the maximum risk of committing a Type I error.	$\alpha = 0.10$
Step III	State the associated test statistic.	$F = MS_C / MS_{Res}$
Step IV	Identify the random sampling distribution of the test statistic when H_0 is true.	$F \underset{=}{d} F(3, 56)$ df if H_0 is true
Step V	State the critical value for rejecting the null hypothesis.	Reject H_0 if $p < 0.10$
Step VI	Calculate the value of the test statistic from the sample data.	$F = 68.820$
Step VII	Analyze the results and make an appropriate inference related to the hypotheses tested.	$p = .000$ Discussion: We cannot assess the cycle effect across the filler since there is a significant interaction. The appropriate hypotheses are: 1) $H_0: \mu_{c1} = \mu_{c2} = \mu_{c3} = \mu_{c4}$ $H_1: \mu_{c1} \neq \mu_{c2} \neq \mu_{c3} \neq \mu_{c4}$ when filler = 1 and 2) $H_0: \mu_{c1} = \mu_{c2} = \mu_{c3} = \mu_{c4}$ $H_1: \mu_{c1} \neq \mu_{c2} \neq \mu_{c3} \neq \mu_{c4}$ when filler = 2

Figure 11.33 (*Continued*)

Step Number	Description	Participant Response
Step I	State the null and research hypotheses.	H_0 : $I_{FC} = 0$ H_1 : $I_{FC} \neq 0$
Step II	State the maximum risk of committing a Type I error.	$\alpha = 0.10$
Step III	State the associated test statistic.	$F = MS_I / MS_{Res}$
Step IV	Identify the random sampling distribution of the test statistic when H_0 is true.	$F \underline{d} F (3, 56)$ df when H_0 is true
Step V	State the critical value for rejecting the null hypothesis.	Reject H_0 if $p < 0.10$
Step VI	Calculate the value of the test statistic from the sample data.	$F = 1.920$
Step VII	Analyze the results and make an appropriate inference related to the hypotheses tested.	$p = .205$ Discussion: Accept H_0. There is no statistical evidence to infer that there is a significant interaction effect between filler and cycle on variability in part density. (This means we can assess each main effect across the other main effect.)

Figure 11.33 (*Continued*)

Step Number	Description	Participant Response
Step I	State the null and research hypotheses.	$H_0: \sigma^2_{F1} = \sigma^2_{F2}$ $H_1: \sigma^2_{F1} \neq \sigma^2_{F2}$
Step II	State the maximum risk of committing a Type I error.	$\alpha = 0.10$
Step III	State the associated test statistic.	$F = MS_I / MS_{Res}$
Step IV	Identify the random sampling distribution of the test statistic when H_0 is true.	$F \underset{d}{=} F(1, 56)$ df when H_0 is true
Step V	State the critical value for rejecting the null hypothesis.	Reject H_0 if $p < 0.10$
Step VI	Calculate the value of the test statistic from the sample data.	$F = 4.898$
Step VII	Analyze the results and make an appropriate inference related to the hypotheses tested.	$p = .058$ Discussion: Reject H_0. We have sufficient statistical evidence to infer that there is a significant difference in variability in part density due to level of filler.

Figure 11.33 *(Continued)*

Step Number	Description	Participant Response
Step I	State the null and research hypotheses.	$H_0: \quad \sigma^2_{C1} = \sigma^2_{C2} = \sigma^2_{C3} = \sigma^2_{C4}$ $H_1: \quad \sigma^2_{C1} \neq \sigma^2_{C2} \neq \sigma^2_{C3} \neq \sigma^2_{C4}$
Step II	State the maximum risk of committing a Type I error.	$\alpha = 0.10$
Step III	State the associated test statistic.	$F = MS_C / MS_{Res}$
Step IV	Identify the random sampling distribution of the test statistic when H_0 is true.	$F \underline{d} F(3, 56)$ df when H_0 is true
Step V	State the critical value for rejecting the null hypothesis.	Reject H_0 if $p < 0.10$
Step VI	Calculate the value of the test statistic from the sample data.	$F = 2.145$
Step VII	Analyze the results and make an appropriate inference related to the hypotheses tested.	$p = .173$ Discussion: Accept H_0. There is no statistical evidence to infer that there is a significant difference in variability in part density due to level of cycle time.

Figure 11.33 (*Continued*)

Case Study 7: Factorial Experiments—A Model III Nested Design

This case study will permit us to explore two additional topics in the design of factorial experiments. One, the issue of nesting treatment factors, was discussed in Chap. 5. The second topic has been more briefly reviewed, and that is the subject of random effects. You will recall that in many experiments, we wish to study a treatment, but not in an incorporated fashion. This is particularly the case when the levels of a treatment effect are not equivalent across levels of a second variable. The assessment of multiple coils of steel, within each of five vendors, would be an example of such a case. In this design example, suggested by Dowdey and Weardon (1983), we will find that we have just such a factor. The example is reprinted with permission of the publisher, John Wiley and Sons Inc.

Statement of the problem: Soda crackers lose their crispness in damp climates unless they are packaged in containers that protect them from humidity. A bakery firm wishes to compare five methods of packaging (including a cardboard box control). Four boxes are selected at random from each method of packaging, assigned numbers, and placed in a chamber in which humidity is maintained at 80 percent for 24 hours. The boxes are opened and three crackers are selected from each box at random to be measured for moisture content. The measurements on the 60 crackers are given in milligrams.

In this example, the first treatment factor, package type (PT), is incorporated and fixed. That is, we wish to compare the five package types across all randomly selected boxes and crackers, and the levels of the treatment were purposively selected by the researcher because they were the specific designs he or she wished to compare. The second treatment variable, box (B), falls into a different category in two ways. First, we would not test the hypothesis that box 1 equals box 2 equals box 3 equals box 4, across all package types. Obviously, the designation of box 1, for example, is meaningless across the package design levels. As a result, we would test for box variability or differences within, rather than across, package types.

The second difference between the two treatments is a bit more subtle. You will recall that the levels of the package type factor were selected by the researcher, due to an interest in comparing these particular packages. This makes the package type treatment a fixed effect. All inferences resulting from the study as related to this treatment would relate only to these five levels.

In the case of box, however, we have a random effect. The levels tested were randomly rather than purposively selected. The purpose of this approach is not to test the treatment levels to determine if the levels tested vary, but to assess whether evidence exists to infer whether significant differences exist in all levels of the research population. In the example of boxes, we are attempting to test whether significant differences may be inferred to exist among all boxes in the population tested, and targeted. The analysis of a random effect changes two conditions. First the statistical hypothesis changes from:

$$H_0 : \mu_1 = \mu_2 = \cdots = \mu_J$$

to:

$$H_0 : \sigma_T^2 = 0$$

which may be expressed as: "There are no significant differences (for our example) in average humidity levels between boxes, within each package type assessed, for all boxes in the population studied."

The second condition which is affected by the inclusion is that the analytical procedure changes. In effect, the appropriate error terms (AET), or denominators of the F test statistics, in mixed (Model III) models are different than for Model I analyses. Again, this topic is too complex for discussion given the scope of this book. Additional information may be obtained by referring to *Experimental Design and Industrial Statistics—Level III* by J. Luftig (1991).

Self-Review Activity 11.8 Figure 11.34 reflects the organization of the independent variables for this experiment. Table 11.7 presents the layout of this design and the statistical hypotheses to be tested and Fig. 11.35 displays the statistical analysis.

Develop the research hypotheses appropriate to this study and provide an analysis of the results and conclusions that might be drawn from this experiment based on the seven-step procedure displayed in Fig. 11.36.

Before we leave this case study, note that there are a number of variations of nested designs for factorial applications. Figure 11.37 provides an overview of some selected designs for three, four, five, and six factors.

Treatment Variable(s)	Method I - Incorporated N - Nested	Classification QN - Quantitative QL - Qualitative	Type F - Fixed R - Random	Number of Levels	Level Description
1. *Package type*	*I*	*QN*	*F*	5	*Control, wax, metal, plastic, metal and plastic*
2. *Box*	*N*	*QN*	*R*	4	*Randomly selected*
3.					
4.					
5.					
6.					
7.					
8.					
Blocked Variables	**Number and Value for Levels Blocked**		**Variables Limited**		**At Level**
1.			1.		
2.			2.		
3.			3.		
4.			4.		

Figure 11.34 Organization of Treatment, Block, and Limited Variables for Case Study 7.

TABLE 11.7 Experimental Design Layout for Case Study 7 (Nested Design, Model III).

Treatment Variable 'A'

Randomly Selected Box (B)

Treatment Variable 'B' Package Type (PT)		1	2	3	4
	Control	73	81	70	67
		75	77	62	69
		77	75	64	62
	Wax	60	64	62	53
		61	67	55	50
		63	62	59	55
	Metal	46	49	52	58
		48	54	62	53
		46	54	56	53
	Plastic	60	49	39	52
		53	42	40	55
		60	52	44	49
	Metal and Plastic	38	45	58	48
		37	47	55	47
		38	49	54	46

Statistical hypotheses to be tested:

1. $H_0: \mu_{T1} = \mu_{T2} = \mu_{T3} = \mu_{T4} = \mu_{T5}$
   $H_1: \mu_{T1} \neq \mu_{T2} \neq \mu_{T3} \neq \mu_{T4} \neq \mu_{T5}$

2. **Within** the Control Group
   $H_0: \sigma_T^2 = 0$
   $H_1: \sigma_T^2 \neq 0$

3. **Within** Wax Paper Boxes
   $H_0: \sigma_T^2 = 0$
   $H_1: \sigma_T^2 \neq 0$

4. **Within** Metal Foil Boxes
   $H_0: \sigma_T^2 = 0$
   $H_1: \sigma_T^2 \neq 0$

5. **Within** Plastic Boxes
   $H_0: \sigma_T^2 = 0$
   $H_1: \sigma_T^2 \neq 0$

6. **Within** Metal Foil and Plastic Boxes
   $H_0: \sigma_T^2 = 0$
   $H_1: \sigma_T^2 \neq 0$

SPSS for MS WINDOWS Release 6.1

```
Summaries of      MOISTURE       Moisture Content
By levels of      PACKAGE        Package Type
                  BOX            Box (Random Effect)
```

Variable	Value	Label	Mean	Std Dev	Cases
For Entire Population			55.8500	10.5023	60
PACKAGE	1.00	Control	71.0000	6.3246	12
BOX	1.00		75.0000	2.0000	3
BOX	2.00		77.6667	3.0551	3
BOX	3.00		65.3333	4.1633	3
BOX	4.00		66.0000	3.6056	3
PACKAGE	2.00	Wax	59.2500	5.0114	12
BOX	1.00		61.3333	1.5275	3
BOX	2.00		64.3333	2.5166	3
BOX	3.00		58.6667	3.5119	3
BOX	4.00		52.6667	2.5166	3
PACKAGE	3.00	Metal	52.5833	4.8140	12
BOX	1.00		46.6667	1.1547	3
BOX	2.00		52.3333	2.8868	3
BOX	3.00		56.6667	5.0332	3
BOX	4.00		54.6667	2.8868	3
PACKAGE	4.00	Plastic	49.5833	7.1536	12
BOX	1.00		57.6667	4.0415	3
BOX	2.00		47.6667	5.1316	3
BOX	3.00		41.0000	2.6458	3
BOX	4.00		52.0000	3.0000	3
PACKAGE	5.00	Metal & Palstic	46.8333	6.7801	12
BOX	1.00		37.6667	.5774	3
BOX	2.00		47.0000	2.0000	3
BOX	3.00		55.6667	2.0817	3
BOX	4.00		47.0000	1.0000	3

Total Cases = 60

Figure 11.35 Statistical Analysis for Self-Review Activity 11.8. Assessing the Effect of Packaging Type on Moisture Level.

******Analysis of Variance -- design 1******

Tests of Significance for MOISTURE

Source of Variation	SS	DF	MS	F	Sig of F
PACKAGE	4467.90	4	1116.97	10.01	.000
BOX WITHIN PACKAGE	1674.42	15	111.63		
BOX WITHIN PACKAGE(1)	352.67	3	117.56	12.87	.000
BOX WITHIN PACKAGE(2)	221.58	3	73.86	8.09	.000
BOX WITHIN PACKAGE(3)	168.25	3	56.08	6.14	.002
BOX WITHIN PACKAGE(4)	445.58	3	148.53	16.26	.000
BOX WITHIN PACKAGE(5)	486.33	3	162.11	17.75	.000
WITHIN CELLS (RESIDUAL)	365.33	40	9.13		

Figure 11.35 (*Continued*)

Step Number	Description	Participant Response
Step I	State the null and research hypotheses.	H_0: $\mu_{PT1} = \mu_{PT2} = \mu_{PT3} = \mu_{PT4} = \mu_{PT5}$ H_1: $\mu_{PT1} \neq \mu_{PT2} \neq \mu_{PT3} \neq \mu_{PT4} \neq \mu_{PT5}$
Step II	State the maximum risk of committing a Type I error.	$\alpha = 0.10$
Step III	State the associated test statistic.	$F = MS_{PT}/MS_B$ *
Step IV	Identify the random sampling distribution of the test statistic when H_0 is true.	$F \underset{=}{d} F\,(4, 15)$ df when H_0 is true
Step V	State the critical value for rejecting the null hypothesis.	Reject H_0 if $p < .10$
Step VI	Calculate the value of the test statistic from the sample data.	$F = 10.01$
Step VII	Analyze the results and make an appropriate inference related to the hypotheses tested.	$p = .000$ Discussion: Reject H_0. We have sufficient statistical evidence to infer that there is a significant difference in moisture content among the five package types.

*Where PT = package type and B = box.

Figure 11.36 Seven-step Procedures for Hypothesis Testing for Self-Review Activity 11.8.

Step Number	Description	Participant Response
Step I	State the null and research hypotheses.	$H_0: \sigma_T^2 = 0$ within PT = 1 (control) $H_1: \sigma_T^2 \neq 0$
Step II	State the maximum risk of committing a Type I error.	$\alpha = 0.10$
Step III	State the associated test statistic.	$F = MS_B / MS_{Res}$ within PT = 1
Step IV	Identify the random sampling distribution of the test statistic when H_0 is true.	$F \underset{=}{d} F(3, 40)$ df when H_0 is true
Step V	State the critical value for rejecting the null hypothesis.	Reject H_0 if $p < .10$
Step VI	Calculate the value of the test statistic from the sample data.	$F = 12.87$
Step VII	Analyze the results and make an appropriate inference related to the hypotheses tested.	$p = .000$ Discussion: Reject H_0. We have sufficient statistical evidence to infer that there is a significant difference in moisture content among boxes when package type 1 (control) is used, for all boxes in the population studied.

Figure 11.36 (*Continued*)

Step Number	Description	Participant Response
Step I	State the null and research hypotheses.	$H_0: \quad \sigma_T^2 = 0$ within PT = 2 (wax paper) $H_1: \quad \sigma_T^2 \neq 0$
Step II	State the maximum risk of committing a Type I error.	$\alpha = 0.10$
Step III	State the associated test statistic.	$F = MS_B / MS_{Res}$ within PT = 2
Step IV	Identify the random sampling distribution of the test statistic when H_0 is true.	$F \underset{=}{d} F(3, 40)$ df when H_0 is true
Step V	State the critical value for rejecting the null hypothesis.	Reject H_0 if $p < 0.10$
Step VI	Calculate the value of the test statistic from the sample data.	$F = 8.09$
Step VII	Analyze the results and make an appropriate inference related to the hypotheses tested.	$p = .000$ Discussion: Reject H_0. We have sufficient statistical evidence to infer that there is a significant difference in moisture content among boxes when package type 2 (wax paper) is used, for all boxes in the population studied.

Figure 11.36 (*Continued*)

Step Number	Description	Participant Response
Step I	State the null and research hypotheses.	$H_0: \sigma_T^2 = 0$ within PT = 3 (metal foil) $H_1: \sigma_T^2 \neq 0$
Step II	State the maximum risk of committing a Type I error.	$\alpha = 0.10$
Step III	State the associated test statistic.	$F = MS_B / MS_{Res}$ within PT = 3
Step IV	Identify the random sampling distribution of the test statistic when H_0 is true.	$F \underset{=}{d} F(3, 40)$ df when H_0 is true
Step V	State the critical value for rejecting the null hypothesis.	Reject H_0 if $p < 0.10$
Step VI	Calculate the value of the test statistic from the sample data.	$F = 6.14$
Step VII	Analyze the results and make an appropriate inference related to the hypotheses tested.	$p = .002$ Discussion: Reject H_0. We have sufficient statistical evidence to infer that there is a significant difference in moisture content among boxes when package type 3 (metal foil) is used, for all boxes in the population studied.

Figure 11.36 (*Continued*)

Step Number	Description	Participant Response
Step I	State the null and research hypotheses.	$H_0: \sigma_T^2 = 0$ within PT = 4 (plastic) $H_1: \sigma_T^2 \neq 0$
Step II	State the maximum risk of committing a Type I error.	$\alpha = 0.10$
Step III	State the associated test statistic.	$F = MS_B / MS_{Res}$ within PT = 4
Step IV	Identify the random sampling distribution of the test statistic when H_0 is true.	$F \underline{d} F (3, 40)$ df when H_0 is true
Step V	State the critical value for rejecting the null hypothesis.	Reject H_0 if $p < 0.10$
Step VI	Calculate the value of the test statistic from the sample data.	$F = 16.26$
Step VII	Analyze the results and make an appropriate inference related to the hypotheses tested.	$p = .000$ Discussion: Reject H_0. We have sufficient statistical evidence to infer that there is a significant difference in moisture content among boxes when package type 4 (plastic) is used, for all boxes in the population studied.

Figure 11.36 (*Continued*)

Step Number	Description	Participant Response
Step I	State the null and research hypotheses.	$H_0: \sigma_T^2 = 0$ within PT = 5 (metal foil and boxes) $H_1: \sigma_T^2 \neq 0$
Step II	State the maximum risk of committing a Type I error.	$\alpha = 0.10$
Step III	State the associated test statistic.	$F = MS_B / MS_{Res}$ within PT = 5
Step IV	Identify the random sampling distribution of the test statistic when H_0 is true.	$F \underline{\underline{d}} F(3, 40)$ df when H_0 is true
Step V	State the critical value for rejecting the null hypothesis.	Reject H_0 if $p < 0.10$
Step VI	Calculate the value of the test statistic from the sample data.	$F = 17.75$
Step VII	Analyze the results and make an appropriate inference related to the hypotheses tested.	$p = .000$ Discussion: Reject H_0. We have sufficient statistical evidence to infer that there is a significant difference in moisture content among boxes when package type 5 (metal foil and plastic) is used, for all boxes in the population studied.

Figure 11.36 (*Continued*)

Chapter Eleven

Sources of Variance	Sums of Squares	Degrees of Freedom	Expectations of Mean Squares	Format of A-Units
A	(5) - CF	m - 1	$\sigma_c^2 + 1\frac{2}{3}\sigma_b^2 + 3\sigma_a^2$	
B in A	(3)+(4)+(5)	m	$\sigma_c^2 + 1\frac{1}{3}\sigma_b^2$	
C in B	(1)+(2)-(3)	m	σ_c^2	
Total	(1)+(2)+(4)-CF	2m-1	4A Three Factors	a b c
A	(7) - CF	m-1	$\sigma_d^2 + 1\frac{1}{2}\sigma_c^2 + 2\frac{1}{2}\sigma_b^2 + 4\sigma_a^2$	
B in A	(5)+(6)-(7)	m	$\sigma_d^2 + 1\frac{1}{6}\sigma_c^2 + 1\frac{1}{2}\sigma_b^2$	
C in B	(3)+(4)-(5)	m	$\sigma_d^2 + 1\frac{1}{6}\sigma_c^2$	
D in C	(1)+(2)-(3)	m	σ_d^2	
Total	(1)+(2)+(4)+(6)-CF	4m-1	4B Four Factors	a b c d
A	(9) - CF	m-1	$\sigma_e^2 + 1\frac{2}{5}\sigma_d^2 + 2\frac{1}{5}\sigma_c^2 + 3\frac{2}{5}\sigma_b^2 + 5\sigma_a^2$	
B in A	(7)+(8)-(9)	m	$\sigma_e^2 + 1\frac{1}{10}\sigma_d^2 + 1\frac{3}{10}\sigma_c^2 + 1\frac{3}{5}\sigma_b^2$	
C in B	(5)+(6)-(7)	m	$\sigma_e^2 + 1\frac{1}{6}\sigma_d^2 + 1\frac{1}{2}\sigma_c^2$	
D in C	(3)+(4)-(5)	m	$\sigma_e^2 + 1\frac{1}{3}\sigma_d^2$	
E in D	(1)+(2)-(3)	m	σ_e^2	
Total	(1)+(2)+(4)+(6)+(8)-CF	5m-1	4C Five Factors	a b c d e
A	(11) - CF	m-1	$\sigma_f^2 + 1\frac{1}{3}\sigma_e^2 + 2\sigma_d^2 + 3\sigma_c^2 + 4\frac{1}{3}\sigma_b^2 + 6\sigma_a^2$	
B in A	(9)+(10)-(11)	m	$\sigma_f^2 + 1\frac{1}{15}\sigma_e^2 + 1\frac{1}{5}\sigma_d^2 + 1\frac{1}{5}\sigma_c^2 + 1\frac{2}{3}\sigma_b^2$	
C in B	(7)+(8)-(9)	m	$\sigma_f^2 + 1\frac{1}{10}\sigma_e^2 + 1\frac{3}{10}\sigma_d^2 + 1\frac{3}{5}\sigma_c^2$	
D in C	(5)+(6)-(7)	m	$\sigma_f^2 + 1\frac{1}{6}\sigma_e^2 + 1\frac{1}{2}\sigma_d^2$	
E in D	(3)+(4)-(5)	m	$\sigma_f^2 + 1\frac{1}{3}\sigma_e^2$	
F in E	(1)+(2)-(3)	m	σ_f^2	
Total	(1)+(2)+(4)+(6)+(8)+(10)-CF	6m-1	4D Six Factors	a b c d e f

Totals Needed to Get Sums of Squares

$(1) = \Sigma a^2$

$(2) = \Sigma b^2$

$(3) = \dfrac{\Sigma(a+b)^2}{2}$

$(4) = \Sigma c^2$

$(5) = \dfrac{\Sigma(a+b+c)^2}{3}$

$(6) = \Sigma c^2$

$(7) = \dfrac{\Sigma(a+b+c+d)^2}{4}$

$(8) = \Sigma e^2$

$(9) = \dfrac{\Sigma(a+b+c+d+e)^2}{5}$

$(10) = \Sigma f^2$

$(11) = \dfrac{\Sigma(a+b+c+d+e+f)^2}{6}$

$CF = \dfrac{(Grand\ Total)^2}{Total\ No.\ of\ Tests}$

Source: T.R. Bainbridge, Staggered Nested Designs for Estimating Variance Components, *Industrial Quality Control*, Vol. 22, No. 1, pp. 12–20, July 1965.

Figure 11.37 Staggered Nested Designs for Three, Four, Five, and Six Factors.

Case Study 8: Fractional Factorial Experiments

The experiments we have reviewed to this point have been relatively simple, because the number of treatment effects have been small. In many instances, we will wish to study a large number of treatment effects simultaneously. Even if all of these factors are examined at only two levels, an experiment with seven main effects will still require 2^7, or 128, runs, or tested treatment conditions. Generally, this will constitute a larger experiment than is economically and practically feasible. Moreover, the utilization of a fractional factorial design makes a complete factorial design unnecessary.

When we use the term *fraction* we are speaking of a carefully prescribed subset of all possible treatment combinations (Natrella, 1963). By measuring only a fraction of the total number of treatment combinations, we can investigate the effects of interest within an economically and practically defensible condition. Obviously, however, a fractional design cannot, by definition, produce all of the information as a full factorial analysis of the same treatment variables. In designing a fractional factorial design, we attempt to "balance" the size of the design with the effects about which we require clear information. There are two aspects of this balancing effort that are important to consider: the size of the design and its resolution.

First, we will discuss design size. Consider a full factorial design for seven treatment effects, each of which is to be studied at two levels. Assuming that each main effect is designated by the letters A, B, C, D, E, F, and G, there are 128 treatment conditions to be analyzed in a full or complete factorial study. It is possible, however, that we may study these effects in 64, or half, of the total number of runs by employing a fractional design referred to as a *one-half replicate*. This design is illustrated in Fig. 11.38.

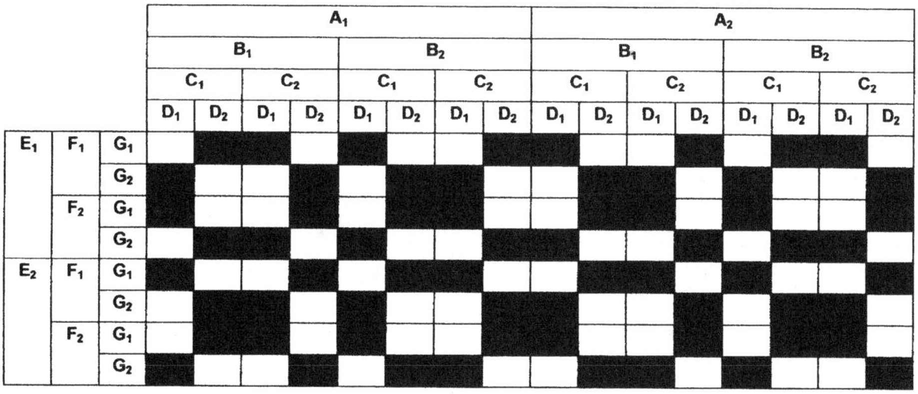

Reprinted with permission from Natrella (1963).

Figure 11.38 One-half Replicate of a 2^7 Factorial.

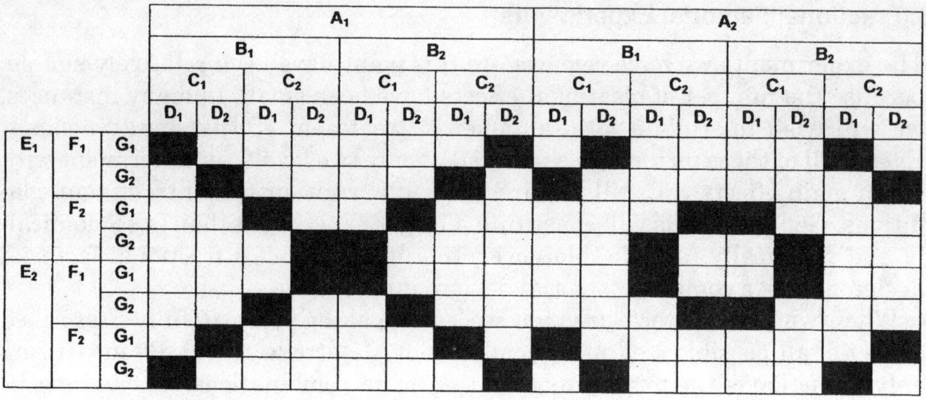

Figure 11.39 One-fourth Replicate of a 2^7 Factorial.

If this number of runs or tests is still prohibitive, then we might consider a one-quarter replicate (Fig. 11.39). This design would permit a study of the seven main effects in only 32 runs.

Alternatively, we could reduce the requirements further to a one-eighth replicate (Fig. 11.40), which would necessitate only sixteen experimental runs or tests. Now, most people who are told that they can substitute 16 tests for 128 tests are bright enough to understand that, as in life, we almost never get "something for nothing." That observation holds true in experimental design as well.

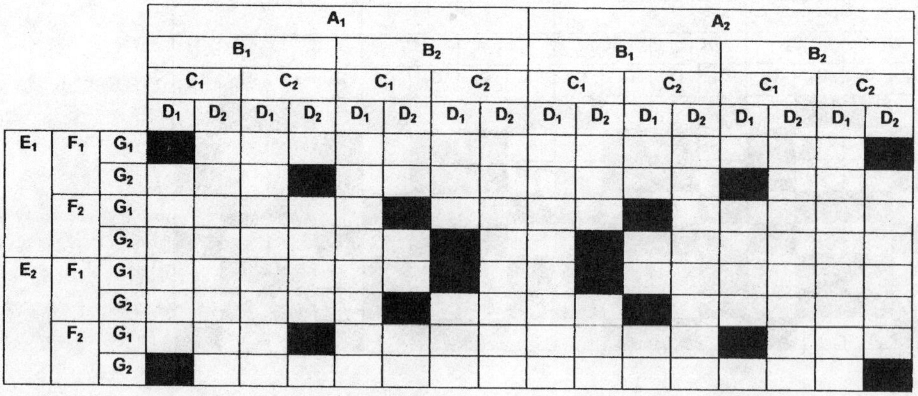

Figure 11.40 One-eighth Replicate of a 2^7 Factorial.

As we reduce the number of runs or treatment combinations tested, we (in general) tend to reduce the resolution of the design. In an effort to reduce the number of tests to be conducted, we are purposively confounding interaction effects with main effects and other interaction effects. This confounding must be purposefully and carefully executed. We do not wish to confound significant interaction effects with main effects. This requires generally two underlying assumptions in our design effort:

1. Higher-order (e.g., three-way, four-way, five-way, etc.) interactions, even if they are significant, are not likely to be important (i.e., explain a great deal of variability).
2. There are a number of two-way interactions that are identifiable in the design process as highly unlikely to show up as significant, according to our understanding of the process.

During the design process, we must be careful to not become overly confident in relying upon these two assumptions, in our desire to reduce the number of runs. Generally speaking, we wish to achieve the lowest number of runs possible, at the highest resolution available. By *resolution,* we are referring to how the main effects and interactions are confounded in the design. These combinations may be described as follows:

1. A resolution III design, the weakest we could employ, confounds main effects with two-way interaction effects. The confounding pattern associated with a fractional design is referred to as the *alias structure*.
2. A resolution IV design does not allow for the confounding of main effects with two-way interactions, but does permit the confounding of main effects with three-way interactions and two-way interactions with other two-way interactions.
3. A resolution V design, which is considered a powerful option for fractional designs, does not allow for main effects to be confounded with two- or three-way interactions, nor does it permit any two-way interaction effect to be confounded with any other two-way interaction. Main effects are confounded with four-way interactions, and two-way interactions are confounded with three-way interactions.

It is absolutely essential that you recognize at this point that the number of runs in a fractional design does not necessarily directly describe the resolution of the design. For example, a one-quarter replicate design may be a resolution III or resolution IV design. The outcome will depend on the way the effects are allocated to the design, the design employed, and the knowledge of the researcher designing the experiment. Obviously, in a given number of runs, we want the highest level of resolution possible within that experimental area.

Many texts are dedicated to various types of fractional designs, or matrices, that may be employed for fractional factorial design efforts. Different researchers and statisticians have published these designs since the 1940s. Not unexpectedly, not all of these matrices are universally applicable or equivalent in their value. In order to illustrate some of these designs, we will examine three fractional factorial applications.

Extreme screening designs

In 1946, Plackett and Burman published a plan for conducting extreme screening designs, also referred to as group screening designs (Juran, Gryna, and Bingham, 1979).

These designs were based upon a number of assumptions:

1. The researcher wishes to identify (or screen out) the critical few main effects from a large number or pool of independent variables.
2. The researcher wishes to perform this analysis with the smallest amount of experimentation possible.
3. All main effects are initially assumed to have the same probability of affecting the criterion measure.
4. The factors to be studied do not interact.

Designing what is sometimes referred to as the "ultimate" screening design, Plackett and Burman created a design methodology employing a resolution III design approach that would allow for the study of K main effects (treatment variables) in $K + 1$ runs or tests. In other words, we should be able to study seven main effects in only eight runs.

Table 11.8 reflects a Plackett-Burman study for our seven-factor analysis. Note that we are studying these treatments in eight rather than 128 runs or tests. This design constitutes a one-sixteenth replicate at a resolution level of III.

Utilizing (typically) an ANOVA, we would generate seven F values (one for each of the treatment variables studied) while expending only eight runs to

TABLE 11.8 A Plackett-Burman Design for N = 8 Runs (Resolution III).

Run	Factor						
	A	B	C	D	E	F	G
1	1	1	1	2	1	2	2
2	2	1	1	1	2	1	2
3	2	2	1	1	1	2	1
4	1	2	2	1	1	1	2
5	2	1	2	2	1	1	1
6	1	2	1	2	2	1	1
7	1	1	2	1	2	2	1
8	2	2	2	2	2	2	2

TABLE 11.9 Alias Structure for the Eight Run Plackett-Burman Design: Two-Way Interaction Confounding Patterns.

Column Effect	Confounded Two-Way Interactions		
A	BF	CD	EG
B	AF	DG	DE
C	AD	BG	EF
D	AC	BE	FG
E	AG	BD	CF
F	AB	CE	DG
G	AE	BC	DF

accomplish the task. Before you become too excited, however, let us review the alias structure for this model. Carefully review Table 11.9.

Let us make certain that you understand the implications of this model. Suppose that, as a result of our experiment, we obtain an F value of 134.34 with an associated p value of 0.000 for the C treatment effect. Note, as related to Table 11.9, that the following two-way interactions are also confounded with the C effect: AD, BG, and EF. This means that the significant F value may, in reality, not be due to the treatment variable C but to one of these three two-way interactions. Now, we note that if the experimental design team had considered this a strong possibility, they would not have selected this design (review the assumptions of the design approach). On the other hand, if the significance of the F value is truly due to BG rather than C, for example, there is no way to determine this condition from this single experiment.

It is due to this effect that no fractional factorial design should ever be conducted without a secondary confirmation experiment. The purpose of this secondary study is to verify that the effects observed are due to the supposed treatment or interaction effects selected for study and not to interactions confounded and identified in the alias structure. Note that, should one of the confounded interactions turn out to be significant and important, it may require a set of successive experiments to identify the true effects. Often, by the time the researcher has completed assessing all of these effects, more experimental conditions have been tested (in total) than would have been used if a higher-resolution fractional design with more than $K+1$ runs had been employed in the first place. In summary, we must be extremely cautious in approving the use of these designs.

Higher-resolution fractional factorial designs

Many statisticians and researchers have published suggested designs, arrays, and matrices for the design of fractional factorial experiments since the 1940s and 1950s. Some of these designs are inherently more powerful than others, in the resolution provided given the number of runs required. Still others tend to be easier to use and modify than comparable matrices of the same size. The key to understanding fractional factorial designs is to recognize that:

1. The best design is one that delivers the highest level of resolution, at the lowest number of runs, with a minimum of inappropriate confounding.
2. The power of any given design or array is inconsequential if it is so difficult to use or modify that the novice researcher must change the research questions to use the design. This is a well-known phenomenon in industrial research. Novice practitioners are taught to use a narrow series of designs, with all variables (for example) limited to two levels. The first time these individuals have to design an experiment with eight or nine treatment variables, some at three or four levels each, with associated interactions between them, the impulse is to try to find an array that matches the design requirements. Finding none, and facing the difficulty of properly modifying the matrices they were taught, they somehow find justification to reduce the number of variables and reduce the incorporated treatments to two levels each. This approach is tantamount to designing the research questions to fit the experimental design; the absolute opposite of what should be done in the conduct of appropriate research.

Having explored these assumptions and conditions, we will review a pair of sample fractional factorial designs of higher (i.e., greater than III) resolution. One of the fractional design categories used for this purpose is that of orthogonal arrays. These arrays, often incorrectly associated with Genichi Taguchi, were developed decades ago. An example of these designs is the L_{16} orthogonal array. The L in the title is a reference to the Latin square, the root or source of this design category. The number 16 tells us that the array consists of 16 runs, or experimental test conditions. If we recognize that the treatments associated with this design are initially set at two levels each, then we could test up to 15 effects at one degree of freedom each using this array. These effects might consist of 15 main effects (not recommended), 10 main effects and 5 two-way interactions, or 5 main effects and 10 two-way interactions. Table 11.10 presents this basic array, with two rows added at the bottom of the array to be utilized in completing the alias structure associated with sample designs to be subsequently developed.

Suppose that we wish to design an experiment that allows us to study eight treatment effects at two levels each. Additionally, let us assume that we wish to study these factors in the context of a resolution IV fractional design, where none of the two-way interactions are considered likely to have a significant and important effect. In this case, we could use the L_{16} orthogonal array to test these effects in the context of a one-sixteenth replicate. Table 11.11 presents one way this array could be utilized for this design.

Note the alias structure completed at the bottom of the array. What are your thoughts regarding this model? How are we protected against significant confounded two-way interactions? How are we not protected against these effects? What options might we have?

TABLE 11.10 The L_{16} Orthogonal Array Fractional Factorial Design.

Effect → Run ↓	1	2	3	4	5	6	7	8	9	10	11	12	13	14	15
1	1	1	1	1	1	1	1	1	1	1	1	1	1	1	1
2	1	1	1	1	1	1	1	2	2	2	2	2	2	2	2
3	1	1	1	2	2	2	2	1	1	1	1	2	2	2	2
4	1	1	1	2	2	2	2	2	2	2	2	1	1	1	1
5	1	2	2	1	1	2	2	1	1	2	2	1	1	2	2
6	1	2	2	1	1	2	2	2	2	1	1	2	2	1	1
7	1	2	2	2	2	1	1	1	1	2	2	2	2	1	1
8	1	2	2	2	2	1	1	2	2	1	1	1	1	2	2
9	2	1	2	1	2	1	2	1	2	1	2	1	2	1	2
10	2	1	2	1	2	1	2	2	1	2	1	2	1	2	1
11	2	1	2	2	1	2	1	1	2	1	2	2	1	2	1
12	2	1	2	2	1	2	1	2	1	2	1	1	2	1	2
13	2	2	1	1	2	2	1	1	2	2	1	1	2	2	1
14	2	2	1	1	2	2	1	2	1	1	2	2	1	1	2
15	2	2	1	2	1	1	2	1	2	2	1	2	1	1	2
16	2	2	1	2	1	1	2	2	1	1	2	1	2	2	1
2 way alias															
3 way alias															

TABLE 11.11 A Resolution IV Fractional Factorial Design.

Effect → Run ↓	A	B	C			F	E			G		H	D		
1	1	1	1	1	1	1	1	1	1	1	1	1	1	1	1
2	1	1	1	1	1	1	1	2	2	2	2	2	2	2	2
3	1	1	1	2	2	2	2	1	1	1	1	2	2	2	2
4	1	1	1	2	2	2	2	2	2	2	2	1	1	1	1
5	1	2	2	1	1	2	2	1	1	2	2	1	1	2	2
6	1	2	2	1	1	2	2	2	2	1	1	2	2	1	1
7	1	2	2	2	2	1	1	1	1	2	2	2	2	1	1
8	1	2	2	2	2	1	1	2	2	1	1	1	1	2	2
9	2	1	2	1	2	1	2	1	2	1	2	1	2	1	2
10	2	1	2	1	2	1	2	2	1	2	1	2	1	2	1
11	2	1	2	2	1	2	1	1	2	1	2	2	1	2	1
12	2	1	2	2	1	2	1	2	1	2	1	1	2	1	2
13	2	2	1	1	2	2	1	1	2	2	1	1	2	2	1
14	2	2	1	1	2	2	1	2	1	1	2	2	1	1	2
15	2	2	1	2	1	1	2	1	2	2	1	2	1	1	2
16	2	2	1	2	1	1	2	2	1	1	2	1	2	2	1
2 way alias			AB CF EG HD	AC BF EH GD	AF BC ED GH		AE BG CH FD	AG BE CD FH		AH BD CE FG			AD BH CG FE		

TABLE 11.12 A Resolution V Fractional Factorial Design.

Effect → Run ↓	A	B	AB	C	AC	BC	DE	D	AD	BD	CE	CD	BE	AE	E
1	1	1	1	1	1	1	1	1	1	1	1	1	1	1	1
2	1	1	1	1	1	1	1	2	2	2	2	2	2	2	2
3	1	1	1	2	2	2	2	1	1	1	1	2	2	2	2
4	1	1	1	2	2	2	2	2	2	2	2	1	1	1	1
5	1	2	2	1	1	2	2	1	1	2	2	1	1	2	2
6	1	2	2	1	1	2	2	2	2	1	1	2	2	1	1
7	1	2	2	2	2	1	1	1	1	2	2	2	2	1	1
8	1	2	2	2	2	1	1	2	2	1	1	1	1	2	2
9	2	1	2	1	2	1	2	1	2	1	2	1	2	1	2
10	2	1	2	1	2	1	2	2	1	2	1	2	1	2	1
11	2	1	2	2	1	2	1	1	2	1	2	2	1	2	1
12	2	1	2	2	1	2	1	2	1	2	1	1	2	1	2
13	2	2	1	1	2	2	1	1	2	2	1	1	2	2	1
14	2	2	1	1	2	2	1	2	1	1	2	2	1	1	2
15	2	2	1	2	1	1	2	1	2	2	1	2	1	1	2
16	2	2	1	2	1	1	2	2	1	1	2	1	2	2	1
2 way alias	All 2 way interactions studied - none confounded with main effects														
3 way alias			CDE		BDE	ADE	ABC		BCE	ACE	ABE	ABD	ACD	BCD	

The L_{16} orthogonal array can also be employed to design resolution V fractional designs. Suppose we wished to test five treatments, at two levels each. For this requirement, the L_{16} orthogonal array allows us to design a one-half replicate, resolution V fractional factorial design. Table 11.12 presents an example of how this design might appear. Compare the alias structure with the previous design.

Once the design is complete and executed, the statistical analysis will be quite similar to the analyses we have already reviewed in this chapter. As an example, suppose we wished to perform the experiment reflected by the variables table shown in Fig. 11.41.

This study was designed to identify the critical variables associated with primer seating force, an end-of-line characteristic that is critical to the customer. The product is used by sportsmen who reload their own ammunition. In this case, higher seating force is preferable.

Note that treatment D, lacquer vendor, is to be studied at four levels. If we were to study these effects in a full factorial design, we would have to test 1024 experimental combinations, or runs. By selecting and modifying an L_{16} orthogonal array, it is possible to analyze these effects in 16 runs, although we will be utilizing a Resolution III design.

Table 11.13 reflects the experimental layout that was employed for this study.

Designing the Industrial Experiment

Treatment Variable(s)	Method I - Incorporated N - Nested	Classification QN - Quantitative QL - Qualitative	Type F - Fixed R - Random	Number of Levels	Level Description
1. Foil paper (A)	I	QL	F	2	Refer to research design proposal
2. Cup burrs (B)	I	QN	F	2	
3. Anvil height (C)	I	QN	F	2	
4. Lacquer vendor (D)	I	QL	F	4	
5. Thickness (E)	I	QN	F	2	
6. Cup O.D. (F)	I	QN	F	2	
7. Pellet weight (G)	I	QN	F	2	
8. Anvil O.D. (H)	I	QN	F	2	
9. Assembly height (I)	I	QN	F	2	

Blocked Variables	Number and Value for Levels Blocked	Variables Limited	At Level
1. None		1. None	
2.		2.	
Two-Way Interactions to be Incorporated		3.	
A x B G x H A x C G x I		4.	

Figure 11.41 Organization of Treatment Effects for a Fractional Factorial Design (Primer testing force study).

TABLE 11.13 A Resolution III Fractional Factorial Design for the Primer Seating Force Study.

Effect → Run ↓	A	D	E	H	I	B	AB	F	G	C	AC	GI	GH
1	1	1	1	1	1	1	1	1	1	1	1	1	1
2	1	2	1	1	1	1	1	2	2	2	2	2	2
3	1	1	1	2	2	2	2	1	1	2	2	2	2
4	1	2	1	2	2	2	2	2	2	1	1	1	1
5	1	3	2	1	1	2	2	1	2	1	1	2	2
6	1	4	2	1	1	2	2	2	1	2	2	1	1
7	1	3	2	2	2	1	1	1	2	2	2	1	1
8	1	4	2	2	2	1	1	2	1	1	1	2	2
9	2	1	2	1	2	1	2	2	2	1	2	1	2
10	2	2	2	1	2	1	2	1	1	2	1	2	1
11	2	1	2	2	1	2	1	2	2	2	1	2	1
12	2	2	2	2	1	2	1	1	1	1	2	1	2
13	2	3	1	1	2	2	1	2	1	1	2	2	1
14	2	4	1	1	2	2	1	1	2	2	1	1	2
15	2	3	1	2	1	1	2	2	1	2	1	1	2
16	2	4	1	2	1	1	2	1	2	1	2	2	1
2 way alias	DE HI DF DG	AE HB FG	AD IB DG FD	AI DB DC	AH EB FC	DH EI DC	DI EH GC	AD DG ED IC	AD DF ED	HD IF BD	HF ID BG	DC HD BD	EC ID BF

ANOVA Table

Source of Variability	Df	S	V	F	rho%
A	1	98.000	98.000	33.362	2.15
D	3	1325.500	441.833	50.411	29.75
E	1	882.000	882.000	300.255	19.86
H	1	40.500	40.500	13.787	0.85
I	1	60.500	60.500	20.596	1.30
B	1	924.500	924.500	314.723	20.82
AB	1	968.000	968.000	329.532	21.80
F	1	40.500	40.500	13.787	0.85
G	1	8.000	8.000	2.723	0.11
C	1	0.500	0.500	0.170	−0.06
AC	1	0.500	0.500	0.170	−0.06
GI	1	18.000	18.000	6.128	0.34
GH	1	12.500	12.500	4.255	0.22
Experimental Error	16	47.000	2.938	2.06	
Total	31	4426.000	142.774		

Figure 11.42 ANOVA for Screening Experiment—L_{16} Orthogonal Array.

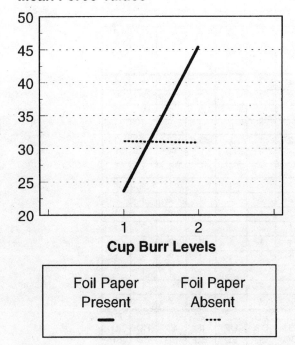

Figure 11.43 Analysis of Interaction Between Foil Paper and Burrs on Mean Force Values.

Designing the Industrial Experiment 263

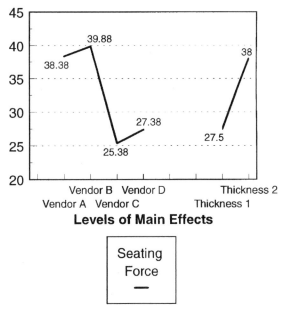

Figure 11.44 Analysis of Critical Main Effects on Mean Force Values.

Two replications were generated for each experimental run tested. The raw data were then processed with a software package specifically designed for the analysis of data associated with orthogonal array designs (ANOVA-TM). The output in Fig. 11.42 is a result of that statistical analysis.

Utilizing the results of the ANOVA, the cell means for the significant and important effects from the analysis were plotted for a visual display of the outcome. Figures 11.43 and 11.44 present the result of this effort.

Self-Review Activity 11.9

1. What conclusions may be drawn from these data?
2. Is it appropriate to make a process change immediately or is there anything else that should be studied at this point?
3. If you believe another study is appropriate, design it.

Additional Experimental Design Approaches

The purpose of this chapter was to provide a series of case studies and examples of the major designs employed in industry. There are a number of other factorial-based designs, such as split-plot and split-split-plot designs that were not reviewed, but are logical extensions of the designs reviewed. However, there are two other research or experimental design "approaches" that warrant a brief discussion: the analysis of covariance and the evolutionary operation technique.

Analysis of covariance

Analysis of covariance (ANCOVA) is a statistical procedure combining ANOVA with regression analysis. Referred to by some authors as a means of increasing the precision of an experimental design (the primary purpose of blocking), ANCOVA can be employed (Kirk, 1968) to:

1. Increase the precision of an experiment
2. Control for extraneous variables that are nonmanipulable or not feasibly manipulated
3. Compare regressions within several groups

In the context of conditions 1 and 2 above, ANCOVA is considered a reasonable option (Wildt and Ahtola, 1978) when:

1. There are one or more extraneous sources of variability that are believed to affect the dependent variable
2. This extraneous source of variation can be measured on a continuous data scale
3. The relationship between the extraneous variables and the dependent variable is known and generally assumed to be linear
4. Direct experimental control of the extraneous variables is not possible or feasible
5. It is possible to obtain a measure of the extraneous variable that is independent of the effects of the treatment variable

While ANCOVA is a popular approach in the behavioral, social, and educational sciences, it is less useful in industry for two primary reasons. First, the assumptions associated with the use of ANCOVA, statistically speaking, are highly restrictive (Sax, 1979):

1. The relationship between the extraneous variable (covariate) and the dependent variable is often unknown and, when known, is not linear.
2. The correlation between the dependent variable and covariate must be high (but not necessarily positive).
3. Initial differences among groups or levels must be randomly distributed.
4. It is assumed that the measure on the covariate is obtained prior to the introduction of the treatment, and therefore cannot be affected by the treatment (which becomes difficult in the event of Type I or II destructive tests for the covariate).

The second reason that ANCOVA is a less attractive option for industrial applications is that the alternative management method of blocking is generally available to the industrial researcher and is a superior option. Some researchers have conducted a significant amount of research comparing the use of ANCOVA with blocking the effects of interest. They concluded that although ANCOVA was slightly more precise, the sensitivity of the procedure against violation of its underlying assumptions did not warrant its use, particularly if blocking was an option. Since this is the general likelihood for the conduct of industrial research, ANCOVA remains an interesting but infrequently utilized statistical procedure.

Evolutionary operation (EVOP) technique

Originally introduced by G. E. P. Box in conjunction with research methods for process industries, EVOP is an alternative to designed experiments (Juran, Gryna, and Bingham, 1979) or a follow-up to designed experiments. The logic behind the EVOP process is that, in many instances, the resources and time required to perform a series of designed experiments may not be feasible. As an alternative, EVOP plans exist that allow for making a series of very small adjustments, over and over, while optimizing or "fine tuning" the process set points or conditions. These authors prefer to consider EVOP a natural outgrowth and valuable secondary approach to the holistic design process. EVOP is not a technique intended to solve major problems quickly or bridge quality gaps in quantum leaps. Its attractiveness lies in the ability to fine-tune the process at little or no additional cost. First, however, the process engineer or researcher must have:

1. Identified the special causes of variability associated with the process
2. Identified the potential optimum conditions associated with the critical first- and second-order process variables
3. Standardized these conditions into the standard operating procedures (SOPs) associated with the day-to-day operation of the process
4. Applied appropriate control charts

Only then may EVOP be of maximum value. Figure 11.45 presents an outline of a general procedure or structure associated with the conduct of EVOP research.

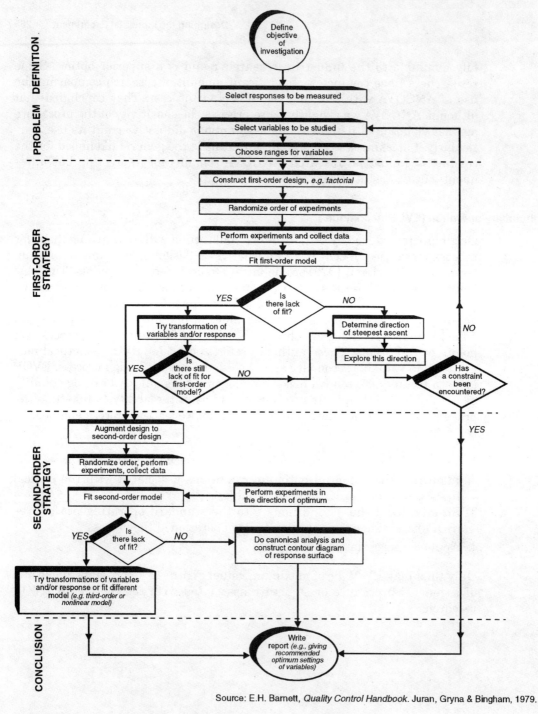

Figure 11.45 Outline of Main Ideas of EVOP and Response Methodology.

Appendix A

A Planning Checklist for Industrial Research

I. Developing the Statement of the Problem

- ☐ A. A comprehensive and complete background data base for the research study exists, originating from a: (✓ one)

    - ☐ Strategic product-market analysis
    - ☐ Technical competitive benchmarking study
    - ☐ Quality improvement effort
    - ☐ Problem-solving effort
    - ☐ Other (specify)_____

- ☐ B. The background data base includes a statement of significance for the problem

- ☐ C. The originating problem has been properly delineated and delimited

- ☐ D. A specific and focused statement of the problem has been developed that is: (all should have a ✓)

    - ☐ Concise
    - ☐ Precise
    - ☐ Testable
    - ☐ Obtainable

II. Classifying the Research Study

The research design to be conducted has been identified as: (✓ one)

- ☐ Agreement research
    - ○ Consensus
    - ○ Instrument/system concordance

- ☐ Descriptive research
    - ○ Status study
    - ○ Longitudinal study
    - ○ Case study
    - ○ Cross-sectional study

- ☐ Relational research
    - ○ Concurrent correlational
    - ○ Predictive correlational
    - ○ Causal comparative

- ☐ Experimental research

III. Operationalizing the Statement of the Problem

- ☐ If the research design is nonexperimental, research questions have been developed; or if the research design is experimental, research hypotheses have been developed

- ☐ The research questions or hypotheses have been evaluated and assessed as: (all should have a ✓)
    - ☐ Clear
    - ☐ Testable
    - ☐ Specific
    - ☐ Simply stated
    - ☐ Consistent with known facts

IV. Designing the Industrial Experiment (all should have a ✓)

- ☐ The target and research populations have been described in clear and specific terms

- ☐ The primary dependent variable(s) and the associated criterion measure(s) has been identified, and is consistent with the research hypothesis(es) to be tested. Secondary dependent variables have also been identified. (The first two columns of checklist attachment A have been completed)

- ☐ All independent variables that may have an affect on the primary or secondary dependent variables have been identified

- ☐ The treatment variables have been identified and classified to be incorporated or nested

- ☐ The known and manipulable nuisance variables have been identified and classified to be blocked or limited

- ☐ The known and nonmanipulable nuisance variables have been identified and an appropriate monitoring plan has been set up for each factor for use during the conduct of the experiment

- ☐ The treatment variables have been correctly categorized as:
  * Qualitative versus quantitative
  * Fixed versus random effects

- ☐ Classification of fixed versus random effects for each treatment variable has been shown to be consistent with the statement of the problem and the research hypothesis(es)

- ☐ For each treatment variable, an appropriate number of levels has been identified. Level selections are appropriately within the operating envelope of the system, set at an extreme level, and have been shown to be nonoverlapping

- ☐ A summary table for independent variable management has been completed (checklist attachment B)

- ☐ A tentative selection has been made corresponding to the type of experimental design which is to be utilized as the basis for the research study (checklist attachment C)

- ☐ An experimental design notification memorandum has been circulated for review, with all appropriate parties signing off on the dependent, treatment, blocked, and limited variables included in the study; and the tentative design selected

ATTACHMENTS FOLLOW

Appendix A

Checklist Attachment A.

A Preliminary Breakdown of Dependent Variables and Criterion Measures for Inclusion in the Experimental Design

Primary Dependent Variable(s)	Criterion Measure(s)	Measurement Scale
1.	1.	
	2.	
	3.	
	4.	
	5.	
2.	1.	
	2.	
	3.	
	4.	
	5.	
3.	1.	
	2.	
	3.	
	4.	
	5.	

Secondary Dependent Variable(s)	Criterion Measure(s)	Measurement Scale
1.	1.	
	2.	
	3.	
2.	1.	
	2.	
	3.	
3.	1.	
	2.	
	3.	

Checklist Attachment B.

Table for the Organization of Treatment, Block, and Limited Variables

Treatment Variable(s)	Method	Classification	Type	Number of Levels	Level Description
	I - Incorporated N - Nested	QN - Quantitative QL - Qualitative	F - Fixed R - Random		
1.					
2.					
3.					
4.					
5.					
6.					
7.					
8.					
9.					
10.					

Block Variables	Number and Value for Levels Blocked	Variables Limited	At Level
1.		1.	
2.		2.	
3.		3.	
4.		4.	

Checklist Attachment C.

A Checksheet for the Identification of the Selected Experimental Design

	Experimental Design	Place a check (✓) next to the design selected
1.	Completely Randomized Design	
2.	Randomized Block Design	
3.	Latin Square design	
4.	Graeco-Latin Square design	
5.	Balanced incomplete block design	
6.	Youden Square balanced incomplete block design	
7.	Partially balanced incomplete block design	
8.	Completely randomized factorial design	
9.	Randomized block factorial design	
10.	Completely randomized nested design	
11.	Completely randomized partially tested nested design	
12.	Split-plot design	
13.	Randomized block completely confounded factorial design	
14.	Latin Square completely confounded factorial design	
15.	Completely randomized fractional factorial design	
	a. Extreme screening design	
	b. Orthogonal Array design (specify L ____)	
	c. High resolution fractional factorial design	
	d. Other	
16.	Randomized block fractional factorial design	
17.	Completely randomized analysis of covariance design	
18.	Randomized block analysis of covariance design	

Checklist Attachment D. A breakdown of measurement scales and data as measured.*

Data Category	Measurement Scale		Description
Qualitative (Discrete, Attribute)	Nominal		Simple classification data; the assignment of numbers for identification purposes. Categorical scales.
	Ordinal	Count	Data corresponding to a frequency measure; such as number of pinholes or scratches per unit.
		Scale	Subjective scale data gathered from measurement or evaluative scales requiring the assignment of a numerical value; such as Likert Scales and Semantic Differential Scales.
		Low Resolution Continuous	Data resulting from low measurement resolution as related to the variability of the process or population measured.
Quantitative (Continuous, Variable)	Interval		Classification according to some continuum characterized by an equality of units, and the ability to sensibly subdivide the units of measure into smaller units.
	Ratio		Interval data, where the ratio between two measures on the scale is meaningful.

* The distinction is made between the nature of the underlying population and the data "delivered" by the instrument.

Checklist Attachment E. Elements and methods associated with the assessment of instrumentation effectiveness.

	Elements			
	Precision	Accuracy	Reliability	Validity
Indices/Metrics	Measurement error due to • Repeatability • Reproducibility	Bias due to Out of calibration conditions	Agreement due to • Internal consistency • Concordance based on • Stability • Equivalence • Internal consistency or homogeneity	Concordance with a standard based on • Content Validity • Predictive Validity • Concurrent Validity • Construct Validity • Face Validity

Methods	• Repeatability and Reproducibility studies
	• Short-term control charts
	• Long-term control charts
		• Short term calibration analyses
		• Long-term control charts
			• Test-retest methods
			• Equivalent forms analyses
			• Split-half and odd-even techniques; Kuder-Richardson analyses
				• Correlative analyses between the measurements obtained and:
				  • Subsequent scores or values
				  • Nonequivalent predicted score or value
				  • Metrics obtained through agreement analyses
				  • Assessments provided by targeted sample

	Quantitative Data		Qualitative Data	

Checklist Attachment F. Responsibility assignment form.

	Tasks to be Accomplished						
Responsibility	Measurement Analysis		Training and Instruction	On-Site Monitoring	Sampling	Testing	Data Analysis
	Continuous	Discrete					
Experimental Design Team Designate							
Statistical Facilitator							
Process Engineer							
Manufacturing Engineer							
Design Engineer							
Technician							
Operator							
First Line Supervisor							
Maintenance Personnel							
Procurement Personnel							
Quality Engineer							

P = Primary Responsibility
S = Secondary Responsibility

Appendix A

Checklist Attachment G. Engineering log.

Engineering Log
Treatment Variables Run Sheet

Date _____   Page ___ of ___

Experimental Run Number _____   Run Reference Number _____

Sponsor _____   Team Leader _____

Factor Number	Label	Treatment Variables	Setting	
			Level	Explanation
1	A			
2	B			
3	C			
4	D			
5	E			
6	F			
7	G			
8	H			
9	I			
10	J			
11	K			
12	L			
13	M			
14	N			
15	O			
Blocked Variables		1. _____ 2. _____	@ _____ @ _____	
Limited Variables		1. _____ 2. _____	@ _____ @ _____	

Checklist Attachment G continued.

Engineering Log
Nuisance (Nonmanipulable) Variables
Monitoring Sheet

Date _____   Page ___ of ___

Experiment _____

Sponsor _____   Team Leader _____

Variable Number	Nuisance Variables	Monitoring Information			
		Checksheet		Control Chart	
		Form Number	Number of Observations/Run	Chart Number	Sample Size (n)
1					
2					
3					
4					
5					
6					
7					
8					
9					
10					
11					
12					
13					
14					
15					
16					
17					
18					

Checklist Attachment G continued.

Engineering Log
Criterion Measure Data Recording Sheet

Date _____   Page ___ of ___

Experiment _____

Experimental Run Number _____   Run Reference Number _____

Sponsor _____   Team Leader _____

| Criterion Measure _____ |
| Responsibility _____ |

Order*	Measure	Order	Measure	Order	Measure
1		7		13	
2		8		14	
3		9		15	
4		10		16	
5		11		17	
6		12		18	

| Criterion Measure _____ |
| Responsibility _____ |

Order	Measure	Order	Measure	Order	Measure
1		7		13	
2		8		14	
3		9		15	
4		10		16	
5		11		17	
6		12		18	

*Order corresponds to test order across runs

Checklist Attachment H.

A Recommended Checklist for the Final Report Resulting from a Research Study

☐ Provide an overview of the background of the problem, and the reason for the conduct of the study.

☐ Present the statement of the problem.

☐ Present the research questions or research hypotheses.

☐ Present the research design employed, including:

☐   The target population.

☐   The dependent variable(s) and criterion measures.

☐   The management plan for the independent variables (use of the independent variable layout form is recommended for this purpose).

☐   The experimental design employed, including:

☐   The type of experimental design

☐   The design layout or array (include a linear graph if an orthogonal array is utilized)

☐   Outcome of the sample size calculations

☐   The sampling plan employed

☐   The statistical hypotheses to be tested

☐ Present the results of the statistical analysis of the data for each criterion measure.

☐ Present the results of the statistical hypothesis tests (the use of Seven-Step Procedure forms in Appendix D are recommended for this purpose).

☐ Present the results and conclusions associated with the tests of the research hypotheses or answer the research questions posed.

☐ Summarize the results of the study as associated with the statement of the problem. Indicate the target population for which the results may be inferred. Describe the populations and conditions for which the results of the study may not be extended.

☐ Present the results of the confirmation experiment, if applicable.

☐ Provide recommendations for:

☐   Further or additional research (verification experiments, descriptive research, secondary full factorial analyses, Evolutionary Operation [EVOP]).

☐ Present the next step in the action plan associated with the background of the problem.

☐ Completed SDCA checklist, if applicable.

Appendix A

Checklist Attachment I. SDCA checklist.

Product design review	Process design review	Materials procurement review	Additional/new training requirements
☐ Changes required to product targets or specifications	☐ Changes required to targets for first and/or second order critical process variables	☐ Action required for new vendor qualification	☐ Modification to training program/system for SOPs required for (specify): _____
☐ Changes required to engineering drawings	☐ Changes required to in-process SOPs	☐ Action required for vendor review	
☐ Changes required to in-process product specifications	☐ Changes required for inspection procedures	☐ Action required for vendor disqualification	☐ Follow-up training required for new control procedures for (specify): _____
☐ Changes required to tooling drawings and/or requirements	☐ New modified test standards required	☐ Action required for changes in targets or specifications for purchased consumable supplies or incoming raw materials	
☐ Concordance analysis to customer requirements necessary	☐ Modifications to existing and/or new control charts required		

Circulation list: _____

Appendix B

Sources of Invalidity for Some Pre-experimental Experimental, and Quasi-experimental designs

Appendix B

	Sources of Internal Invalidity							
	History	Maturation	Testing	Instrumentation	Regression	Selection	Mortality	Interaction of Selection and Maturation
Pre-Experimental Designs:								
1. One-Shot Case Study X O	—	—				—	—	
2. One-Group Pretest-Posttest Design O X O	—	—	—	—	?	+	+	—

3. Static-Group Comparison
    X O
        O
	+	?	+	+	+	—	—	—

True Experimental Designs:								

4. Pretest-Posttest Control Group Design
    R O X O
    R O   O
	+	+	+	+	+	+	+	+

5. Solomon Four-Group Design
    R O X O
    R O   O
    R   X O
    R   O
	+	+	+	+	+	+	+	+

6. Posttest-Only Control Group Design
    R X O
    R   O
	+	+	+	+	+	+	+	+

Legend: + Controls for effects
            — Does not control for effects
            ? May control for effects under certain conditions
            ---- Nonequivalent groups

Campbell, Donald T. and Julian C. Stanley, *Experimental and Quasi-Experimental Designs for Research*. Copyright © 1963 by Houghton Mifflin Company. Used with permission.

Sources of Invalidity

	Internal								External			
	History	Maturation	Testing	Instrumentation	Regression	Selection	Mortality	Interaction of Selection and Maturation	Interaction of Testing and X	Interaction of Selection and X	Reactive Arrangement	Multiple-X Interference
Quasi-Experimental Designs:												
7. Time Series O O O OXO O O O	−	+	+	?	+	+	+	+	−	?	?	
8. Equivalent Time Samples Design $X_1O\ X_0O\ X_1O\ X_0O$, etc.	+	+	+	+	+	+	+	+	−	?	−	−
9. Equivalent Materials Samples Design $M_aX_1O\ M_bX_0O\ M_cX_1O\ M_dX_0O$, etc.	+	+	+	+	+	+	+	+	−	?	?	−
10. Nonequivalent Control Group Design O....X....O O O	+	+	+	+	?	+	+	−	−	?	?	
11. Counterbalanced Designs $X_1O\ X_2O\ X_3O\ X_4O$ $X_2O\ X_4O\ X_1O\ X_3O$ $X_3O\ X_1O\ X_4O\ X_2O$ $X_4O\ X_3O\ X_2O\ X_1O$	+	+	+	+	+	+	+	?	?	?	?	−
12. Separate-Sample Pretest—Posttest Design R O (X) R X O	−	−	+	?	+	+	−	−	+	+	+	
12a. R O (X) R............X....O R O (X) R X O	+	−	+	?	+	+	−	+	+	+	+	
12b. R O_1 R O_2 (X) R X O_3	−	+	+	?	+	+	−	?	+	+	+	
12c. R O_1 X O_2 R X O_3	−	−	+	?	+	+	+	−	+	+	+	

Legend: + Controls for effects
        − Does not control for effects
        ? May control for effects under certain conditions
       ---- Nonequivalent groups

Campbell, Donald T. and Julian C. Stanley, *Experimental and Quasi-Experimental Designs for Research*. Copyright © 1963 by Houghton Mifflin Company. Used with permission.

Appendix B

	Sources of Invalidity											
	Internal								External			
	History	Maturation	Testing	Instrumentation	Regression	Selection	Mortality	Interaction of Selection and Maturation	Interaction of Testing and X	Interaction of Selection and X	Reactive Arrangement	Multiple-X Interference

Quasi-Experimental Designs: (continued)

13. Separate-Sample Pretest-Posttest Control Group Design

```
R   O   (X)
R       X    O
R   O
R            O
```

History	Maturation	Testing	Instrumentation	Regression	Selection	Mortality	Int. Sel×Mat	Int. Test×X	Int. Sel×X	Reactive	Mult-X
+	+	+	+	+	+	+	−	+	+	+	

13a.

```
       ⎧ R   O   (X)
       ⎪ R        X    O
    R' ⎨ R   O   (X)
       ⎪ R        X    O
       ⎪ R   O   (X)
       ⎩ R        X    O
       ⎧ R   O
       ⎪ R              O
    R' ⎨ R   O
       ⎪ R              O
       ⎪ R   O
       ⎩ R              O
```

| + | + | + | + | + | + | + | + | + | + | + | |

14. Multiple Time-Series

```
O  O  OXO  O  O
O  O   O   O  O
```

| + | + | + | + | + | + | + | + | − | − | ? | |

15. Institutional Cycle Design

```
Class  A      X   O₁
Class  B₁  RO₂ X  O₃
Class  B₂  R   X  O₄  ⎫
Class  C          O₅  ⎬ X
                      ⎭
*Gen. Pop. Con. Cl.  B  O₆  ⎫
*Gen. Pop.           C  O₇  ⎬
                            ⎭
O₂ < O₁
O₃ < O₄
O₂ < O₃
O₂ < O₄
O₆ = O₇
O₂ᵧ = O₂₀
```

History	Mat.	Test.	Instr.	Regr.	Sel.	Mort.	Int SM	Int TX	Int SX	React.	Mult-X
+	−	+	+	?	−	?		+	?	+	
−	−	−	?	?	+	+		−	?	+	
−	−	+	?	?	+	?		+	?	?	
+					−						

16. Regression Discontinuity

| + | + | + | ? | + | + | ? | + | + | − | + | + |

Legend: + Controls for effects
         − Does not control for effects
         ? May control for effects under certain conditions
         ---- Nonequivalent groups

Campbell, Donald T. and Julian C. Stanley, *Experimental and Quasi-Experimental Designs for Research*. Copyright © 1963 by Houghton Mifflin Company. Used with permission.

Appendix C

Selected Tables of Critical Values

Appendix C

Table C-1. Table of areas under the *Normal* curve.

z	Area	z	Area	z	Area	z	Area
.00	.00000	.41	.15910	.82	.29389	1.23	.39065
.01	.00399	.42	.16276	.83	.29673	1.24	.39251
.02	.00798	.43	.16640	.84	.29955	1.25	.39435
.03	.01197	.44	.17003	.85	.30234	1.26	.39617
.04	.01596	.45	.17365	.86	.30511	1.27	.39796
.05	.01994	.46	.17724	.87	.30785	1.28	.39973
.06	.02393	.47	.18082	.88	.31057	1.29	.40148
.07	.02791	.48	.18439	.89	.31327	1.30	.40320
.08	.03188	.49	.18794	.90	.31594	1.31	.40490
.09	.03586	.50	.19146	.91	.31859	1.32	.40658
.10	.03983	.51	.19498	.92	.32121	1.33	.40824
.11	.04380	.52	.19847	.93	.32382	1.34	.40988
.12	.04776	.53	.20195	.94	.32639	1.35	.41149
.13	.05172	.54	.20540	.95	.32895	1.36	.41309
.14	.05567	.55	.20884	.96	.33147	1.37	.41466
.15	.05962	.56	.21226	.97	.33398	1.38	.41621
.16	.06356	.57	.21566	.98	.33646	1.39	.41774
.17	.06750	.58	.21904	.99	.33891	1.40	.41924
.18	.07143	.59	.22241	1.00	.34135	1.41	.42073
.19	.07535	.60	.22575	1.01	.34375	1.42	.42220
.20	.07926	.61	.22907	1.02	.34614	1.43	.42364
.21	.08317	.62	.23237	1.03	.34850	1.44	.42507
.22	.08707	.63	.23565	1.04	.35083	1.45	.42647
.23	.09096	.64	.23892	1.05	.35314	1.46	.42786
.24	.09484	.65	.24216	1.06	.35543	1.47	.42922
.25	.09871	.66	.24537	1.07	.35769	1.48	.43056
.26	.10257	.67	.24857	1.08	.35993	1.49	.43189
.27	.10642	.68	.25175	1.09	.36214	1.50	.43319
.28	.11026	.69	.25490	1.10	.36433	1.51	.43448
.29	.11409	.70	.25804	1.11	.36650	1.52	.43575
.30	.11791	.71	.26115	1.12	.36864	1.53	.43699
.31	.12172	.72	.26424	1.13	.37076	1.54	.43822
.32	.12552	.73	.26731	1.14	.37286	1.55	.43943
.33	.12930	.74	.27035	1.15	.37493	1.56	.44062
.34	.13307	.75	.27337	1.16	.37698	1.57	.44179
.35	.13683	.76	.27637	1.17	.37900	1.58	.44295
.36	.14058	.77	.27935	1.18	.38100	1.59	.44408
.37	.14431	.78	.28231	1.19	.38298	1.60	.44520
.38	.14803	.79	.28524	1.20	.38493	1.61	.44630
.39	.15173	.80	.28815	1.21	.38686	1.62	.44738
.40	.15542	.81	.29103	1.22	.38877	1.63	.44845

Table C-1. Table of areas under the *Normal* curve.

z	Area	z	Area	z	Area	z	Area
1.64	.44950	2.05	.47982	2.46	.49305	2.87	.49795
1.65	.45053	2.06	.48030	2.47	.49324	2.88	.49801
1.66	.45154	2.07	.48077	2.48	.49343	2.89	.49807
1.67	.45254	2.08	.48124	2.49	.49361	2.90	.49813
1.68	.45352	2.09	.48169	2.50	.49379	2.91	.49819
1.69	.45449	2.10	.48214	2.51	.49396	2.92	.49825
1.70	.45544	2.11	.48257	2.52	.49413	2.93	.49831
1.71	.45637	2.12	.48300	2.53	.49430	2.94	.49836
1.72	.45728	2.13	.48341	2.54	.49446	2.95	.49841
1.73	.45819	2.14	.48382	2.55	.49461	2.96	.49846
1.74	.45907	2.15	.48422	2.56	.49477	2.97	.49851
1.75	.45994	2.16	.48461	2.57	.49492	2.98	.49856
1.76	.46080	2.17	.48500	2.58	.49506	2.99	.49861
1.77	.46164	2.18	.48537	2.59	.49520	3.00	.49865
1.78	.46246	2.19	.48574	2.60	.49534	3.01	.49869
1.79	.46327	2.20	.48610	2.61	.49547	3.02	.49874
1.80	.46407	2.21	.48645	2.62	.49560	3.03	.49878
1.81	.46485	2.22	.48679	2.63	.49573	3.04	.49882
1.82	.46562	2.23	.48713	2.64	.49586	3.05	.49886
1.83	.46638	2.24	.48746	2.65	.49598	3.06	.49889
1.84	.46712	2.25	.48778	2.66	.49609	3.07	.49893
1.85	.46784	2.26	.48809	2.67	.49621	3.08	.49897
1.86	.46856	2.27	.48840	2.68	.49632	3.09	.49900
1.87	.46926	2.28	.48870	2.69	.49643	3.10	.49903
1.88	.46995	2.29	.48899	2.70	.49653	3.11	.49906
1.89	.47062	2.30	.48928	2.71	.49664	3.12	.49910
1.90	.47128	2.31	.48956	2.72	.49674	3.13	.49913
1.91	.47193	2.32	.48983	2.73	.49683	3.14	.49916
1.92	.47257	2.33	.49010	2.74	.49693	3.15	.49918
1.93	.47320	2.34	.49036	2.75	.49702	3.16	.49921
1.94	.47381	2.35	.49061	2.76	.49711	3.17	.49924
1.95	.47441	2.36	.49086	2.77	.49720	3.18	.49926
1.96	.47500	2.37	.49111	2.78	.49728	3.19	.49929
1.97	.47558	2.38	.49134	2.79	.49737	3.20	.49931
1.98	.47615	2.39	.49158	2.80	.49745	3.21	.49934
1.99	.47671	2.40	.49180	2.81	.49752	3.22	.49936
2.00	.47725	2.41	.49202	2.82	.49760	3.23	.49938
2.01	.47779	2.42	.49224	2.83	.49767	3.24	.49940
2.02	.47831	2.43	.49245	2.84	.49774	3.25	.49942
2.03	.47882	2.44	.49266	2.85	.49781	3.26	.49944
2.04	.47933	2.45	.49286	2.86	.49788	3.27	.49946

Appendix C

Table C-1. Table of areas under the *Normal* curve.

z	Area	z	Area
3.28	.49948	3.69	.49989
3.29	.49950	3.70	.49989
3.30	.49952	3.71	.49990
3.31	.49953	3.72	.49990
3.32	.49955	3.73	.49990
3.33	.49957	3.74	.49991
3.34	.49958	3.75	.49991
3.35	.49960	3.76	.49992
3.36	.49961	3.77	.49992
3.37	.49962	3.78	.49992
3.38	.49964	3.79	.49992
3.39	.49965	3.80	.49993
3.40	.49966	3.81	.49993
3.41	.49968	3.82	.49993
3.42	.49969	3.83	.49994
3.43	.49970	3.84	.49994
3.44	.49971	3.85	.49994
3.45	.49972	3.86	.49994
3.46	.49973	3.87	.49995
3.47	.49974	3.88	.49995
3.48	.49975	3.89	.49995
3.49	.49976	3.90	.49995
3.50	.49977	3.91	.49995
3.51	.49978	3.92	.49996
3.52	.49978	3.93	.49996
3.53	.49979	3.94	.49996
3.54	.49980	3.95	.49996
3.55	.49981	3.96	.49996
3.56	.49982	3.97	.49996
3.57	.49982	3.98	.49997
3.58	.49983	3.99	.49997
3.59	.49984	4.00	.49997
3.60	.49984		
3.61	.49985		
3.62	.49985		
3.63	.49986		
3.64	.49986		
3.65	.49987		
3.66	.49987		
3.67	.49988		
3.68	.49988		

Notes: *z* Column values: Calculated *z* scores. **Area** column values: Area under the normal curve from μ to *z*.

Table C-2. Percentage points, Student's t distribution (upper-tail probabilities).

(One-tail probabilities)

α \ ν	0.40	0.25	0.10	0.05	0.025	0.01	0.005	0.0005
1	0.325	1.000	3.078	6.314	12.706	31.821	63.657	636.619
2	0.289	0.816	1.886	2.920	4.303	6.965	9.925	31.598
3	0.277	0.765	1.638	2.353	3.182	4.541	5.841	12.941
4	0.271	0.741	1.533	2.132	2.776	3.747	4.604	8.610
5	0.267	0.727	1.476	2.015	2.571	3.365	4.032	6.859
6	0.265	0.718	1.440	1.943	2.447	3.143	3.707	5.959
7	0.263	0.711	1.415	1.895	2.365	2.998	3.499	5.405
8	0.262	0.706	1.397	1.860	2.306	2.896	3.355	5.041
9	0.261	0.703	1.383	1.833	2.262	2.821	3.250	4.781
10	0.260	0.700	1.372	1.812	2.228	2.764	3.169	4.587
11	0.260	0.697	1.363	1.796	2.201	2.718	3.106	4.437
12	0.259	0.695	1.356	1.782	2.179	2.681	3.055	4.318
13	0.259	0.694	1.350	1.771	2.160	2.650	3.012	4.221
14	0.258	0.692	1.345	1.761	2.145	2.624	2.977	4.140
15	0.258	0.691	1.341	1.753	2.131	2.602	2.947	4.073
16	0.258	0.690	1.337	1.746	2.120	2.583	2.921	4.015
17	0.257	0.689	1.333	1.740	2.110	2.567	2.898	3.965
18	0.257	0.688	1.330	1.734	2.101	2.552	2.878	3.922
19	0.257	0.688	1.328	1.729	2.093	2.539	2.861	3.883
20	0.257	0.687	1.325	1.725	2.086	2.528	2.845	3.850
21	0.257	0.686	1.323	1.721	2.080	2.518	2.831	3.819
22	0.256	0.686	1.321	1.717	2.074	2.508	2.819	3.792
23	0.256	0.685	1.319	1.714	2.069	2.500	2.807	3.767
24	0.256	0.685	1.318	1.711	2.064	2.492	2.797	3.745
25	0.256	0.684	1.316	1.708	2.060	2.485	2.787	3.725
26	0.256	0.684	1.315	1.706	2.056	2.479	2.779	3.707
27	0.256	0.684	1.314	1.703	2.052	2.473	2.771	3.690
28	0.256	0.683	1.313	1.701	2.048	2.467	2.763	3.674
29	0.256	0.683	1.311	1.699	2.045	2.462	2.756	3.659
30	0.256	0.683	1.310	1.697	2.042	2.457	2.750	3.646
40	0.255	0.681	1.303	1.684	2.021	2.423	2.704	3.551
60	0.254	0.679	1.296	1.671	2.000	2.390	2.660	3.460
120	0.254	0.677	1.289	1.658	1.980	2.358	2.617	3.373
∞	0.253	0.674	1.282	1.645	1.960	2.326	2.576	3.291
	0.80	0.50	0.20	0.10	0.05	0.02	0.01	0.001

(Two-tail probability)

Table C-3. Percentage points, F distribution. $\alpha = 0.10$ (upper tail).

ν_2 \ ν_1	1	2	3	4	5	6	7	8	9
1	39.86	49.50	53.59	55.83	57.24	58.20	58.91	59.44	59.86
2	8.53	9.00	9.16	9.24	9.29	9.33	9.35	9.37	9.38
3	5.54	5.46	5.39	5.34	5.31	5.28	5.27	5.25	5.24
4	4.54	4.32	4.19	4.11	4.05	4.01	3.98	3.95	3.94
5	4.06	3.78	3.62	3.52	3.45	3.40	3.37	3.34	3.32
6	3.78	3.46	3.29	3.18	3.11	3.05	3.01	2.98	2.96
7	3.59	3.26	3.07	2.96	2.88	2.83	2.78	2.75	2.72
8	3.46	3.11	2.92	2.81	2.73	2.67	2.62	2.59	2.56
9	3.36	3.01	2.81	2.69	2.61	2.55	2.51	2.47	2.44
10	3.29	2.92	2.73	2.61	2.52	2.46	2.41	2.38	2.35
11	3.23	2.86	2.66	2.54	2.45	2.39	2.34	2.30	2.27
12	3.18	2.81	2.61	2.48	2.39	2.33	2.28	2.24	2.21
13	3.14	2.76	2.56	2.43	2.35	2.28	2.23	2.20	2.16
14	3.10	2.73	2.52	2.39	2.31	2.24	2.19	2.15	2.12
15	3.07	2.70	2.49	2.36	2.27	2.21	2.16	2.12	2.09
16	3.05	2.67	2.46	2.33	2.24	2.18	2.13	2.09	2.06
17	3.03	2.64	2.44	2.31	2.22	2.15	2.10	2.06	2.03
18	3.01	2.62	2.42	2.29	2.20	2.13	2.08	2.04	2.00
19	2.99	2.61	2.40	2.27	2.18	2.11	2.06	2.02	1.98
20	2.97	2.59	2.38	2.25	2.16	2.09	2.04	2.00	1.96
21	2.96	2.57	2.36	2.23	2.14	2.08	2.02	1.98	1.95
22	2.95	2.56	2.35	2.22	2.13	2.06	2.01	1.97	1.93
23	2.94	2.55	2.34	2.21	2.11	2.05	1.99	1.95	1.92
24	2.93	2.54	2.33	2.19	2.10	2.04	1.98	1.94	1.91
25	2.92	2.53	2.32	2.18	2.09	2.02	1.97	1.93	1.89
26	2.91	2.52	2.31	2.17	2.08	2.01	1.96	1.92	1.88
27	2.90	2.51	2.30	2.17	2.07	2.00	1.95	1.91	1.87
28	2.89	2.50	2.29	2.16	2.06	2.00	1.94	1.90	1.87
29	2.89	2.50	2.28	2.15	2.06	1.99	1.93	1.89	1.86
30	2.88	2.49	2.28	2.14	2.05	1.98	1.93	1.88	1.85
40	2.84	2.44	2.23	2.09	2.00	1.93	1.87	1.83	1.79
60	2.79	2.39	2.18	2.04	1.95	1.87	1.82	1.77	1.74
120	2.75	2.35	2.13	1.99	1.90	1.82	1.77	1.72	1.68
∞	2.71	2.30	2.08	1.94	1.85	1.77	1.72	1.67	1.63

Table C-3. Percentage points, *F* distribution. $\alpha = 0.10$ (upper tail).

v_2 \ v_1	10	12	15	20	24	30	40	60	120	∞
1	60.19	60.71	61.22	61.74	62.00	62.26	62.53	62.79	64.06	63.33
2	9.39	9.41	9.42	9.44	9.45	9.46	9.47	9.47	9.48	9.49
3	5.23	5.22	5.20	5.18	5.18	5.17	5.16	5.15	5.14	5.13
4	3.92	3.90	3.87	3.84	3.83	3.82	3.80	3.79	3.78	3.76
5	3.30	3.27	3.24	3.21	3.19	3.17	3.16	3.14	3.12	3.10
6	2.94	2.90	2.87	2.84	2.82	2.80	2.78	2.76	2.74	2.72
7	2.70	2.67	2.63	2.59	2.58	2.56	2.54	2.51	2.49	2.47
8	2.54	2.50	2.46	2.42	2.40	2.38	2.36	2.34	2.32	2.29
9	2.42	2.38	2.34	2.30	2.28	2.25	2.24	2.21	2.18	2.15
10	2.32	2.28	2.24	2.20	2.18	2.16	2.13	2.11	2.08	2.06
11	2.25	2.21	2.17	2.12	2.10	2.08	2.05	2.03	2.00	1.97
12	2.19	2.15	2.10	2.06	2.04	2.01	1.99	1.96	1.93	1.90
13	2.14	2.10	2.05	2.01	1.98	1.96	1.93	1.90	1.88	1.85
14	2.10	2.05	2.01	1.96	1.94	1.91	1.89	1.86	1.83	1.80
15	2.06	2.02	1.97	1.92	1.90	1.87	1.85	1.82	1.79	1.76
16	2.03	1.99	1.94	1.89	1.87	1.84	1.81	1.78	1.75	1.72
17	2.00	1.96	1.91	1.86	1.84	1.81	1.78	1.75	1.72	1.69
18	1.98	1.93	1.89	1.84	1.81	1.78	1.75	1.72	1.69	1.66
19	1.96	1.91	1.86	1.81	1.79	1.76	1.73	1.70	1.67	1.63
20	1.94	1.89	1.84	1.79	1.77	1.74	1.71	1.68	1.64	1.61
21	1.92	1.87	1.83	1.78	1.75	1.72	1.69	1.66	1.62	1.59
22	1.90	1.86	1.81	1.76	1.73	1.70	1.67	1.64	1.60	1.57
23	1.89	1.84	1.80	1.74	1.72	1.69	1.66	1.62	1.59	1.55
24	1.88	1.83	1.78	1.73	1.70	1.67	1.64	1.61	1.57	1.53
25	1.87	1.82	1.77	1.72	1.69	1.66	1.63	1.59	1.56	1.52
26	1.86	1.81	1.76	1.71	1.68	1.65	1.61	1.58	1.54	1.50
27	1.85	1.80	1.75	1.70	1.67	1.64	1.60	1.57	1.53	1.49
28	1.84	1.79	1.74	1.69	1.66	1.63	1.59	1.56	1.52	1.48
29	1.83	1.78	1.73	1.68	1.65	1.62	1.58	1.55	1.51	1.47
30	1.82	1.77	1.72	1.67	1.64	1.61	1.57	1.54	1.50	1.46
40	1.76	1.71	1.66	1.61	1.57	1.54	1.51	1.47	1.42	1.38
60	1.71	1.66	1.60	1.54	1.51	1.48	1.44	1.40	1.35	1.29
120	1.65	1.60	1.55	1.48	1.45	1.41	1.37	1.32	1.26	1.19
∞	1.60	1.55	1.49	1.42	1.38	1.34	1.30	1.24	1.17	1.00

Table C-3. Percentage points, F distribution. $\alpha = 0.05$ (upper tail).

v_2 \ v_1	1	2	3	4	5	6	7	8	9
1	161.40	199.50	215.70	224.60	230.20	234.00	236.80	238.90	240.50
2	18.51	19.00	19.16	19.25	19.30	19.33	19.35	19.37	19.38
3	10.13	9.55	9.28	8.12	9.01	8.94	8.89	8.85	8.81
4	7.71	6.94	6.29	6.39	6.26	6.16	6.08	6.04	6.00
5	6.61	5.79	5.41	5.19	5.05	4.94	4.88	4.82	4.77
6	5.99	5.14	4.76	4.53	4.39	4.28	4.21	4.15	4.10
7	5.59	4.74	4.35	4.12	3.97	3.87	3.79	3.73	3.68
8	5.32	4.46	4.07	3.84	3.69	3.58	3.50	3.44	3.39
9	5.12	4.26	3.86	3.63	3.48	3.37	3.29	3.23	3.18
10	4.96	4.10	3.71	3.48	3.33	3.22	3.14	3.07	3.02
11	4.84	3.98	3.59	3.36	3.20	3.09	3.01	2.95	2.90
12	4.75	3.89	3.49	3.26	3.11	3.00	2.91	2.85	2.80
13	4.67	3.81	3.41	3.18	3.03	2.92	2.83	2.77	2.71
14	4.60	3.74	3.34	3.11	2.96	2.85	2.76	2.70	2.65
15	4.54	3.68	3.29	3.06	2.90	2.79	2.71	2.64	2.59
16	4.49	3.63	3.24	3.01	2.85	2.74	2.66	2.59	2.54
17	4.45	3.59	3.20	2.96	2.81	2.70	2.61	2.55	2.49
18	4.41	3.55	3.16	2.93	2.77	2.66	2.58	2.51	2.46
19	4.38	3.52	3.13	2.90	2.74	2.63	2.54	2.48	2.42
20	4.35	3.49	3.10	2.87	2.71	2.60	2.51	2.45	2.39
21	4.32	3.47	3.07	2.84	2.68	2.57	2.49	2.42	2.37
22	4.30	3.44	3.05	2.82	2.66	2.55	2.46	2.40	2.34
23	4.28	3.42	3.03	2.80	2.64	2.53	2.44	2.37	2.32
24	4.26	3.40	3.01	2.78	2.62	2.51	2.42	2.36	2.30
25	4.24	3.39	2.99	2.76	2.60	2.49	2.40	2.34	2.28
26	4.23	3.37	2.98	2.74	2.59	2.47	2.39	2.32	2.27
27	4.21	3.35	2.96	2.73	2.57	2.46	2.37	2.31	2.25
28	4.20	3.34	2.95	2.71	2.56	2.45	2.36	2.29	2.24
29	4.18	3.33	2.93	2.70	2.55	2.43	2.35	2.28	2.22
30	4.17	3.32	2.92	2.69	2.53	2.42	2.33	2.27	2.21
40	4.08	3.23	2.84	2.61	2.45	2.34	2.25	2.18	2.12
60	4.00	3.15	2.76	2.53	2.37	2.25	2.17	2.10	2.04
120	3.92	3.07	2.68	2.45	2.29	2.17	2.09	2.02	1.96
∞	3.84	3.00	2.60	2.37	2.21	2.10	2.01	1.94	1.88

Table C-3. Percentage points, F distribution. $\alpha = 0.05$ (upper tail).

v_1 \ v_2	10	12	15	20	24	30	40	60	120	∞
1	241.90	243.90	245.90	248.00	249.10	250.10	251.10	252.20	253.30	254.30
2	19.40	19.41	19.43	19.45	19.45	19.46	19.47	19.48	19.49	19.50
3	8.79	8.74	8.70	8.68	8.64	8.62	8.59	8.57	8.55	8.53
4	5.96	5.91	5.86	5.80	5.77	5.75	5.72	5.69	5.66	5.63
5	4.74	4.68	4.62	4.56	4.53	4.50	4.46	4.43	4.40	4.36
6	4.06	4.00	3.94	3.87	3.84	3.81	3.77	3.74	3.70	3.67
7	3.64	3.57	3.51	3.44	3.41	3.38	3.34	3.30	3.27	3.23
8	3.35	3.28	3.22	3.15	3.12	3.08	3.04	3.01	2.97	2.93
9	3.14	3.07	3.01	2.94	2.90	2.86	2.83	2.79	2.75	2.71
10	2.98	2.91	2.85	2.77	2.74	2.70	2.66	2.62	2.58	2.54
11	2.85	2.79	2.72	2.65	2.61	2.57	2.53	2.49	2.45	2.40
12	2.75	2.69	2.62	2.54	2.51	2.47	2.43	2.38	2.34	2.30
13	2.67	2.60	2.53	2.46	2.42	2.38	2.34	2.30	2.25	2.21
14	2.60	2.53	2.46	2.39	2.35	2.31	2.27	2.22	2.18	2.13
15	2.54	2.48	2.40	2.33	2.29	2.25	2.20	2.16	2.11	2.07
16	2.49	2.42	2.35	2.28	2.24	2.19	2.15	2.11	2.06	2.01
17	2.45	2.38	2.31	2.23	2.19	2.15	2.10	2.06	2.01	1.96
18	2.41	2.34	2.27	2.19	2.15	2.11	2.06	2.02	1.97	1.92
19	2.38	2.31	2.23	2.16	2.11	2.07	2.03	1.98	1.93	1.88
20	2.35	2.28	2.20	2.12	2.08	2.04	1.99	1.95	1.90	1.84
21	2.32	2.25	2.18	2.10	2.05	2.01	1.95	1.92	1.87	1.81
22	2.30	2.23	2.15	2.07	2.03	1.98	1.94	1.89	1.84	1.78
23	2.27	2.20	2.13	2.05	2.01	1.96	1.91	1.86	1.81	1.76
24	2.25	2.18	2.11	2.03	1.98	1.94	1.89	1.84	1.79	1.73
25	2.24	2.16	2.09	2.01	1.96	1.92	1.87	1.82	1.77	1.71
26	2.22	2.15	2.07	1.99	1.95	1.90	1.85	1.80	1.75	1.69
27	2.20	2.13	2.06	1.97	1.93	1.88	1.84	1.79	1.73	1.67
28	2.19	2.12	2.04	1.96	1.91	1.87	1.82	1.77	1.71	1.65
29	2.18	2.10	2.03	1.94	1.90	1.85	1.81	1.75	1.70	1.64
30	2.16	2.09	2.01	1.93	1.89	1.84	1.79	1.74	1.68	1.62
40	2.08	2.00	1.92	1.84	1.79	1.74	1.69	1.64	1.58	1.51
60	1.99	1.92	1.84	1.75	1.70	1.65	1.59	1.53	1.47	1.39
120	1.91	1.83	1.75	1.66	1.61	1.55	1.50	1.43	1.35	1.25
∞	1.83	1.75	1.67	1.57	1.52	1.46	1.39	1.32	1.22	1.00

Table C-3. Percentage points, F distribution. $\alpha = 0.025$ (upper tail).

v_2 \ v_1	1	2	3	4	5	6	7	8	9
1	647.80	799.50	864.2	899.60	921.80	937.10	848.20	956.70	963.30
2	38.51	39.00	39.17	39.25	39.30	39.33	39.36	39.37	39.39
3	17.44	16.04	15.44	15.10	14.88	14.73	14.62	14.54	14.47
4	12.22	10.65	9.98	9.60	9.36	9.20	9.07	8.98	8.90
5	10.01	8.43	7.76	7.39	7.15	6.98	6.85	6.76	6.68
6	8.81	7.26	6.60	6.23	5.99	5.82	5.70	5.60	5.52
7	8.07	6.54	5.89	5.52	5.29	5.12	4.99	4.90	4.82
8	7.57	6.06	5.42	5.05	4.82	4.65	4.54	4.43	4.36
9	7.21	5.71	5.08	4.72	4.48	4.32	4.20	4.10	4.03
10	6.94	5.46	4.83	4.47	4.24	4.07	3.95	3.85	3.78
11	6.72	5.26	4.63	4.28	4.04	3.88	3.76	3.66	3.59
12	6.55	5.10	4.47	4.12	3.89	3.73	3.61	3.51	3.44
13	6.41	4.97	4.35	4.00	3.77	3.60	3.48	3.39	3.31
14	6.30	4.86	4.24	3.89	3.66	3.50	3.38	3.29	3.21
15	6.20	4.77	4.15	3.80	3.58	3.41	3.29	3.20	3.12
16	6.12	4.69	4.08	3.73	3.50	3.34	3.22	3.12	3.05
17	6.04	4.62	4.01	3.66	3.44	3.28	3.16	3.06	2.98
18	5.98	4.56	3.95	3.61	3.38	3.22	3.10	3.01	2.93
19	5.92	4.51	3.90	3.56	3.33	3.17	3.05	2.96	2.88
20	5.87	4.46	3.86	3.51	3.29	3.13	3.01	2.91	2.84
21	5.83	4.42	3.82	3.48	3.25	3.09	2.97	2.87	2.80
22	5.79	4.38	3.78	3.44	3.22	3.05	2.93	2.84	2.76
23	5.75	4.35	3.75	3.41	3.18	3.02	2.90	2.81	2.73
24	5.72	4.32	3.72	3.38	3.15	2.99	2.87	2.78	2.70
25	5.69	4.29	3.69	3.35	3.13	2.97	2.85	2.75	2.68
26	5.66	4.27	3.67	3.33	3.10	2.94	2.82	2.73	2.65
27	5.63	4.24	3.65	3.31	3.08	2.92	2.80	2.71	2.63
28	5.61	4.22	3.63	3.29	3.06	2.90	2.78	2.69	2.61
29	5.59	4.20	3.61	3.27	3.04	2.88	2.76	2.67	2.59
30	5.57	4.18	3.59	3.25	3.03	2.87	2.75	2.65	2.57
40	5.42	4.05	3.46	3.13	2.90	2.74	2.62	2.53	2.45
60	5.29	3.93	3.34	3.01	2.79	2.63	2.51	2.41	2.33
120	5.15	3.80	3.23	2.89	2.67	2.52	2.39	2.30	2.22
∞	5.02	3.69	3.12	2.79	2.57	2.41	2.29	2.19	2.11

Table C-3. Percentage points, *F* distribution. $\alpha = 0.025$ (upper tail).

v_2 \ v_1	10	12	15	20	24	30	40	60	120	∞
1	968.60	976.70	984.90	993.10	997.20	1001.00	1006.00	1010.00	1014.00	1018.00
2	39.40	39.41	39.43	39.45	39.46	39.46	39.47	39.48	39.49	39.50
3	14.42	14.34	14.25	14.17	14.12	14.08	14.04	13.99	13.95	13.90
4	8.84	8.75	8.66	8.56	8.51	8.46	8.41	8.36	8.31	8.26
5	6.62	6.52	6.43	6.33	6.28	6.23	6.18	6.12	6.07	6.02
6	5.46	5.37	5.27	5.17	5.12	5.07	5.01	4.96	4.90	4.85
7	4.76	4.67	4.57	4.47	4.42	4.36	4.31	4.25	4.20	4.14
8	4.30	4.20	4.10	4.00	3.95	3.89	3.84	3.78	3.73	3.67
9	3.96	3.87	3.77	3.67	3.61	3.56	3.51	3.45	3.39	3.33
10	3.72	3.62	3.52	3.42	3.37	3.31	3.26	3.20	3.14	3.08
11	3.53	3.43	3.33	3.23	3.17	3.12	3.06	3.00	2.94	2.86
12	3.37	3.28	3.18	3.07	3.02	2.96	2.91	2.85	2.79	2.72
13	3.25	3.15	3.05	2.94	2.89	2.84	2.78	2.72	2.66	2.60
14	3.15	3.05	2.95	2.84	2.79	2.73	2.67	2.61	2.55	2.49
15	3.06	2.96	2.86	2.76	2.70	2.64	2.59	2.52	2.46	2.40
16	2.99	2.89	2.79	2.68	2.63	2.57	2.51	2.45	2.38	2.32
17	2.92	2.82	2.72	2.62	2.56	2.50	2.44	2.38	2.32	2.25
18	2.87	2.77	2.67	2.56	2.50	2.44	2.38	2.32	2.26	2.19
19	2.82	2.72	2.62	2.51	2.45	2.39	2.33	2.27	2.20	2.13
20	2.77	2.68	2.57	2.46	2.41	2.35	2.29	2.22	2.16	2.09
21	2.73	2.64	2.53	2.42	2.37	2.31	2.25	2.18	2.11	2.04
22	2.70	2.60	2.50	2.39	2.33	2.27	2.21	2.14	2.08	2.00
23	2.67	2.57	2.47	2.36	2.30	2.24	2.18	2.11	2.04	1.97
24	2.64	2.54	2.44	2.33	2.27	2.21	2.15	2.08	2.01	1.94
25	2.61	2.51	2.41	2.30	2.24	2.18	2.12	2.05	1.98	1.91
26	2.59	2.49	2.39	2.28	2.22	2.16	2.09	2.03	1.95	1.88
27	2.57	2.47	2.36	2.25	2.19	2.13	2.07	2.00	1.93	1.85
28	2.55	2.45	2.34	2.23	2.17	2.11	2.05	1.98	1.91	1.83
29	2.53	2.43	2.32	2.21	2.15	2.09	2.03	1.96	1.89	1.81
30	2.51	2.41	2.31	2.20	2.14	2.07	2.01	1.94	1.87	1.79
40	2.39	2.29	2.18	2.07	2.01	1.94	1.88	1.80	1.72	1.64
60	2.27	2.17	2.06	1.94	1.88	1.82	1.74	1.67	1.58	1.48
120	2.16	2.05	1.94	1.82	1.76	1.69	1.61	1.53	1.43	1.31
∞	2.05	1.94	1.83	1.71	1.64	1.57	1.48	1.39	1.27	1.00

Table C-3. Percentage points, F distribution. $\alpha = 0.001$ (upper tail).

v_2 \ v_1	1	2	3	4	5	6	7	8	9
1	4052.00	4999.50	5403.00	5625.00	5764.00	5859.00	5928.00	5982.00	6022.00
2	98.50	99.00	99.17	99.25	99.30	99.33	99.36	99.37	99.39
3	34.12	30.82	29.46	28.71	28.24	27.91	27.67	27.49	27.35
4	21.20	18.00	16.69	15.98	15.52	15.21	14.98	14.80	14.66
5	16.26	13.27	12.06	11.39	10.97	10.67	10.46	10.29	10.16
6	13.75	10.92	9.78	9.15	8.75	8.47	8.26	8.10	7.98
7	12.25	9.55	8.45	7.85	8.46	7.19	6.99	6.84	6.72
8	11.26	8.65	7.59	7.01	6.63	6.37	6.18	6.03	5.91
9	10.56	8.02	6.99	6.42	6.06	5.80	5.61	5.47	5.35
10	10.04	7.56	6.55	5.99	5.64	5.39	5.20	5.06	4.94
11	9.65	7.21	6.22	5.67	5.32	5.07	4.89	4.74	4.63
12	9.33	6.93	5.95	5.41	5.06	4.82	4.64	4.50	4.39
13	9.07	6.70	5.74	5.21	4.86	4.62	4.44	4.30	4.19
14	8.86	6.51	5.56	5.04	4.69	4.46	4.28	4.14	4.03
15	8.68	6.36	5.42	4.89	4.56	4.32	4.14	4.00	3.89
16	8.53	6.23	5.29	4.77	4.44	4.20	4.03	3.89	3.78
17	8.40	6.11	5.18	4.67	4.34	4.10	3.93	3.79	3.68
18	8.29	6.01	5.09	4.58	4.25	4.01	3.84	3.71	3.60
19	8.18	5.93	5.01	4.50	4.17	3.94	3.77	3.63	3.52
20	8.10	5.85	4.94	4.43	4.10	3.87	3.70	3.56	3.46
21	8.02	5.78	4.87	4.37	4.04	3.81	3.64	3.51	3.40
22	7.95	5.72	4.82	4.31	3.99	3.76	3.59	3.45	3.35
23	7.88	5.66	4.76	4.26	3.94	3.71	3.54	3.41	3.30
24	7.82	5.61	4.72	4.22	3.90	3.67	3.50	3.36	3.26
25	7.77	5.57	4.68	4.18	3.85	3.63	3.46	3.32	3.22
26	7.72	5.53	4.64	4.14	3.82	3.59	3.42	3.29	3.18
27	7.68	5.49	4.60	4.11	3.78	3.56	3.39	3.26	3.15
28	7.64	5.45	4.57	4.07	3.75	3.53	3.36	3.23	3.12
29	7.60	5.42	4.54	4.04	3.73	3.50	3.33	3.20	3.09
30	7.56	5.39	4.51	4.02	3.70	3.47	3.30	3.17	3.07
40	7.31	5.18	4.31	3.83	3.51	3.29	3.12	2.99	2.89
60	7.08	4.98	4.13	3.65	3.34	3.12	2.95	2.82	2.72
120	6.85	4.79	3.95	3.48	3.17	2.96	2.79	2.66	2.56
∞	6.63	4.61	3.78	3.32	3.02	2.80	2.64	2.51	2.41

Table C-3. Percentage points, F distribution. $\alpha = 0.001$ (upper tail).

v_1 v_2	10	12	15	20	24	30	40	60	120	∞
1	6056.00	6106.00	6157.00	6209.00	6235.00	6261.00	6287.00	6313.00	6339.00	6366.00
2	99.40	99.42	99.43	99.45	99.46	99.47	99.47	99.48	99.49	99.50
3	27.23	27.05	26.87	26.69	26.60	26.50	26.41	26.32	26.22	26.13
4	14.55	14.37	14.20	14.02	13.93	13.84	13.75	13.65	13.56	13.46
5	10.05	9.89	9.72	9.55	9.47	9.38	9.29	9.20	9.11	9.02
6	7.87	7.72	7.53	7.40	7.31	7.23	7.14	7.06	6.97	6.88
7	6.62	6.47	6.31	6.16	6.07	5.99	5.91	5.82	5.74	5.65
8	5.81	5.67	5.52	5.35	5.28	5.20	5.12	5.03	4.95	4.86
9	5.26	5.11	4.96	4.81	4.73	4.65	4.57	4.48	4.40	4.31
10	4.85	4.71	4.56	4.41	4.33	4.25	4.17	4.08	4.00	3.91
11	4.54	4.40	4.25	4.10	4.02	3.94	3.86	3.78	3.69	3.60
12	4.30	4.16	4.01	3.83	3.78	3.70	3.62	3.54	3.45	3.36
13	4.10	3.96	3.82	3.66	3.59	3.51	3.43	3.34	3.25	3.17
14	3.94	3.80	3.66	3.51	3.43	3.35	3.27	3.18	3.09	3.00
15	3.80	3.67	3.52	3.37	3.29	3.21	3.13	3.05	2.96	2.87
16	3.69	3.55	3.41	3.26	3.18	3.10	3.02	2.93	2.84	2.75
17	3.59	3.46	3.31	3.16	3.08	3.00	2.92	2.83	2.75	2.65
18	3.51	3.37	3.23	3.03	3.00	2.92	2.84	2.75	2.66	2.57
19	3.43	3.30	3.15	3.00	2.92	2.84	2.70	2.67	2.58	2.49
20	3.37	3.23	3.09	2.94	2.86	2.78	2.69	2.61	2.52	2.42
21	3.31	3.17	3.03	2.88	2.80	2.72	2.64	2.55	2.46	2.36
22	3.26	3.12	2.98	2.83	2.75	2.67	2.58	2.50	2.40	2.31
23	3.21	3.07	2.93	2.78	2.70	2.62	2.54	2.45	2.35	2.26
24	3.17	3.03	2.89	2.74	2.66	2.58	2.49	2.40	2.31	2.21
25	3.13	2.99	2.85	2.70	2.62	2.54	2.45	2.36	2.27	2.17
26	3.09	2.96	2.81	2.66	2.58	2.50	2.42	2.33	2.23	2.13
27	3.06	2.93	2.78	2.63	2.55	2.47	2.38	2.29	2.20	2.10
28	3.03	2.90	2.75	2.60	2.52	2.44	2.35	2.26	2.17	2.06
29	3.00	2.87	2.73	2.57	2.49	2.41	2.33	2.23	2.14	2.03
30	2.93	2.84	2.70	2.55	2.47	2.39	2.30	2.21	2.11	2.01
40	2.80	2.63	2.52	2.37	2.29	2.20	2.11	2.02	1.92	1.80
60	2.63	2.50	2.35	2.20	2.12	2.03	1.94	1.84	1.73	1.60
120	2.47	2.34	2.19	2.03	1.95	1.86	1.76	1.66	1.53	1.38
∞	2.32	2.18	2.04	1.88	1.79	1.70	1.59	1.47	1.32	1.00

Appendix D

Seven-Step Procedures for Hypothesis Testing

Step Number	Description	Participant Response
Step I	State the null and research hypotheses.	H_0 : H_1 :
Step II	State the maximum risk of committing a Type I error.	$\alpha =$
Step III	State the associated test statistic.	
Step IV	Identify the random sampling distribution of the test statistic when H_0 is true.	
Step V	State the critical value for rejecting the null hypothesis.	Reject H_0 if p <
Step VI	Calculate the value of the test statistic from the sample data.	
Step VII	Analyze the results and make an appropriate inference related to the hypotheses tested.	p = Discussion:

Seven-Step Procedure for Hypothesis Testing

Step Number	Description	Participant Response
Step I	State the null and research hypotheses.	H_0 : H_1 :
Step II	State the maximum risk of committing a Type I error.	$\alpha =$
Step III	State the associated test statistic.	
Step IV	Identify the random sampling distribution of the test statistic when H_0 is true.	
Step V	State the critical value for rejecting the null hypothesis.	Reject H_0 if p <
Step VI	Calculate the value of the test statistic from the sample data.	
Step VII	Analyze the results and make an appropriate inference related to the hypotheses tested.	p = Discussion:

Appendix D

Step Number	Description	Participant Response
Step I	State the null and research hypotheses.	H_0 : H_1 :
Step II	State the maximum risk of committing a Type I error.	$\alpha =$
Step III	State the associated test statistic.	
Step IV	Identify the random sampling distribution of the test statistic when H_0 is true.	
Step V	State the critical value for rejecting the null hypothesis.	Reject H_0 if p <
Step VI	Calculate the value of the test statistic from the sample data.	
Step VII	Analyze the results and make an appropriate inference related to the hypotheses tested.	p = Discussion:

Step Number	Description	Participant Response
Step I	State the null and research hypotheses.	H_0 : H_1 :
Step II	State the maximum risk of committing a Type I error.	$\alpha =$
Step III	State the associated test statistic.	
Step IV	Identify the random sampling distribution of the test statistic when H_0 is true.	
Step V	State the critical value for rejecting the null hypothesis.	Reject H_0 if p <
Step VI	Calculate the value of the test statistic from the sample data.	
Step VII	Analyze the results and make an appropriate inference related to the hypotheses tested.	p = Discussion:

Appendix D

Step Number	Description	Participant Response
Step I	State the null and research hypotheses.	H_0 : H_1 :
Step II	State the maximum risk of committing a Type I error.	$\alpha =$
Step III	State the associated test statistic.	
Step IV	Identify the random sampling distribution of the test statistic when H_0 is true.	
Step V	State the critical value for rejecting the null hypothesis.	Reject H_0 if $p <$
Step VI	Calculate the value of the test statistic from the sample data.	
Step VII	Analyze the results and make an appropriate inference related to the hypotheses tested.	$p =$ Discussion:

Appendix E

Answers to Self-Review Activities

Self-Review Activity 1.1

1. Since we are evaluating historical data (a 3-year history of time to failure and time for replacement) to compare MTBF for the three current suppliers, the type of research used is analytical research (historical).
2. *Statement of the problem:* The purpose of this study is to determine whether the life of slitter knives varies for knives purchased from our three current suppliers. The knives of interest for this study are all knives used on this machine for the last 3 years. Life will be measured on the basis of time to failure and time to replacement.
3. *Research hypothesis:* There is no significant difference in slitter knife life (as measured by time to failure and time to replacement) for our three current suppliers.

Self-Review Activity 1.2

Research hypothesis: There is no significant difference in humidity incursion (as measured with the ASTM-B123 test procedure for five different packaging methods for the CookieSnack and YummyMunch lines in the Southeastern region).

Self-Review Activity 3.1

Conclusion: Assuming a significance level of 0.10, since $p = .000$ (<0.10), reject the null hypothesis. We have sufficient statistical evidence to infer that there is a significant difference in end-of-line flatness of beer bottle neck labels produced at the Fayetteville plant between the four paper suppliers. The next step would be a post-hoc analysis to determine which suppliers lead to the flattest labels (and with less variability).

Self-Review Activity 4.1

The one-group pretest-posttest group design fails to protect us against several types of threats to internal validity (numbers correspond to the list of threats to internal validity in Fig. 4.4):

1. *History.* Since no control group is used, it is possible for something to have changed in the process between the two sets of measurements or observations, something that may not have been affected by the treatment at all.
2. *Selection.* If randomization is not used, any difference seen at two points in time may be due to a change resulting from the test itself rather than the factors studied.
3. *Testing.* If randomization is not used, any difference seen at two points in time may be due to a change resulting from the test itself rather than the factors studied.
4. *Instrumentation.* If randomization is not used, any difference seen at two points in time may be due to a change in the measurement system rather than the factors studied.
5. *Maturation.* If no control group or randomization is employed, differences due to maturation (i.e., fatigue, run-in, etc.) may be mistaken for a change due to the factors studied.
6. *Interaction of selection and maturation.* Since there is no control group and the samples are not selected randomly, it is not possible to determine whether another set of experimental units would perform differently in terms of maturation (fatigue, run-in, etc.).

Self-Review Activity 4.2

1. The pretest-posttest control group design protects against all sources of threats to internal validity including the following:
    a. *History.* A potential threat (addressed by this design) would be that maintenance costs increase due to time related factors such as an increase in fuel costs, increase in the hourly rate charged by maintenance employees, or damage to cars due to lack of oil changes or some other factors besides the factor tested (fuel grade).
    b. *Maturation.* As cars age, they require more maintenance.
    c. *Selection.* If cars are not selected at random we may have chosen one model for gasohol and another model for medium-grade gasoline (currently used). One model may have higher maintenance costs.

d. *Mortality.* Cars are taken out of the fleet due to age, mileage, wrecks, or an employee change.
2. The pretest-posttest control group design protects the researcher against extraneous variables becoming confounded with the treatment in that the two groups of cars (for medium-grade gas and gasohol) are selected at random (which prevents confounding with factors such as age, mileage, and model). The control group protects against confounding effects over time in that if there is a time related factor, it will affect both groups and be detected by the posttest.

Self-Review Activity 4.3

The Solomon four-group design is even stronger than the pretest-posttest control group design and the posttest-only control group design because it combines the strengths of the tests (randomization, a control group, pretest) to protect against all threats to internal validity but eliminates the weaknesses. By exposing one treatment and one control group to a pretest and one treatment and one control group to no pretest, we can determine whether or not the pretested experimental units respond to the treatment effect in the same way that experimental units not pretested will respond (which is important to assuring external validity).

Self-Review Activity 5.1

1. *Statement of the problem:* The purpose of this study is to determine what factors in our process affect sheet cleanliness. The target population is all units produced through the cleaning process on both shifts. The criterion measure is sheet cleanliness as measured in the lab.
2. *Research hypotheses:*
   a. There is no significant difference in sheet cleanliness between the three vendors.
   b. There is no significant difference in sheet cleanliness between the two threading procedures.
   c. There is no significant difference in sheet cleanliness between two levels of temperature.
   d. There is no significant difference in sheet cleanliness between two levels of concentration.
   e. There is no significant difference in sheet cleanliness between two levels of line speed.
   f. There is no significant difference in sheet cleanliness between two levels of squeegee pressure.
3. Independent variables to be included: vendor, threading procedure, temperature, concentration, line speed, and squeegee pressure.

	Treatment Variable(s)	Method I - Incorporated N - Nested	Classification QN - Quantitative QL - Qualitative	Type F - Fixed R - Random	Number of Levels	Level Description
1.	Vendor	I	QL	F	3	Ajax, Quality, Foamo
2.	Threading procedure	I	QL	F	2	1,2
3.	Temperature	I	QN	F	2	145, 165°
4.	Concentration	I	QN	F	2	35, 45%
5.	Line speed	I	QN	F	2	400, 450 FQM
6.	Squeegee pressure	I	QN	F	2	10, 15 PSJ
7.						
8.						
9.						
10.						
	Blocked Variables	**Number and Value for Levels Blocked**		**Variables Limited**		**At Level**
1.	Lab technician	2 - Bob and Kathy		1.		
2.	Shift	2 - Day and night		2.		
3.				3.		
4.				4.		

Figure E.1 Sample Table for the Organization of Treatment, Block, and Limited Variables.

    4. See Fig. E.1.

Case Study 1—Self-Review Activity 11.1

1. There were two research hypotheses posed:
   a. Vendor 4 (our current coolant supplier) provides coolant that, when used on the no. 6 machine under standard operating conditions, produces lower deviation from target product than would be produced under the same conditions with coolant supplied by any of the three new proposed suppliers.
   b. There is no significant difference in the variability of deviation from target values for product manufactured with coolant supplied from our current or three new proposed suppliers.
   Each hypothesis is addressed below:
   a. The statistical analysis shows that we have sufficient statistical evidence to infer that the first research hypothesis is not true. Vendor 4 (the current vendor) does not produce lower deviation from target product than would be produced under the same conditions with coolant supplied by the three proposed suppliers. Vendor 4 does provide significantly lower deviation from target product than supplier 3, but there is no significant difference between suppliers 1, 2, and 4.
   b. The data show that we do not have sufficient statistical evidence to infer any difference in the variability of deviation from target product for the four different suppliers.
2. *Conclusion:* Investigate other factors such as cost and delivery performance for coolant suppliers 1, 2, and 4 for use on machine no. 6 at the Albion Plant. As far as product quality is concerned, there is no significant difference between suppliers 1, 2, and 4 with respect to central tendency. Since the general definition of quality is the reduction of variability around a customer-defined target, an analysis to determine whether there is any signif-

icant difference in variability between suppliers 1, 2, and 4 would be necessary before selecting the "best" supplier. Such an analysis is beyond the scope of this book. Given our analysis of central tendency, if there is no significant difference in variability, the sourcing decision between those three suppliers should be based on other factors (cost, delivery performance, financial health of the supplier company, etc.).

Case Study 2—Self-Review Activity 11.2

1. The research hypothesis (there is no significant difference in the average overtime due to unplanned breakdowns associated with the current vibration system and that associated with the proposed vibration/heat transfer system) was rejected on the basis of the statistical analysis. The data showed that we have sufficient statistical evidence to infer that there is a significant difference in overtime hours associated with the system used. In fact, method 2 (the current vibration method) yields a statistically significantly higher average overtime than the vibration/heat transfer method.

2. Since the data show that the vibration method (our current method) results in a higher average overtime from unplanned breakdowns than the vibration/heat transfer method, the next step would be to weigh the additional cost of the new method versus the cost associated with the difference in overtime hours (and resulting loss of production, missed shipments, cost of repair, etc.). The data indicate that the vibration/heat transfer method could save, on average, 28 hours of overtime per quarter resulting from unplanned breakdowns.

3. Yes, the plant effect is important. If the plant effect had been ignored (refer back to Fig. 11.16), $t = -.67$ with $p = .512$, so we would have concluded that there was no significant difference between the two methods which is contrary to our findings when the plant effect is blocked. Matching was necessary to remove the significant plant effect and appropriately compare the two preventive maintenance methods. If five plants had been randomly selected to test the new method and five plants had been randomly selected to test the current method, plant to plant differences would have inflated the error term and we probably would have concluded (incorrectly) that there was no significant difference between the two methods because such a sampling plan assumes that there is no plant effect which, in this case, is not a valid assumption.

Case Study 3—Self-Review Activity 11.3

1. The research hypothesis (there is no significant difference in average yield between the hybrids tested) was rejected on the basis of the data analysis presented in case study 3. We found that we have sufficient statistical evidence to infer that there is a difference in yield for the four hybrids.

2. The post-hoc analysis indicates that there is no significant difference in average yield between hybrids 1, 2, and 3. These three hybrids give an overall average yield of 64.21 bushels per acre, which is statistically significantly higher than the average yield from hybrid 4 (57.60 bushels per acre).
3. On the basis of the post-hoc analysis, hybrid 4 (RC-3) should not be used since it gives a statistically significantly (and important) lower average yield than hybrids 1, 2, or 3. Since there is no significant difference in average yield between hybrids 1, 2, and 3, the decision should be to use FR-11, BCM, or DBC (but not RC-3). The selection of one of these three hybrids as "best" should be based on the determination of any significant difference between the hybrids with respect to yield dispersion (as mentioned in case study 1). If there is no significant difference in variability, then the choice between hybrids 1, 2, or 3 can be based on other factors such as cost, delivery, difficulty to plant and/or harvest, etc.

Case Study 4—Self-Review Activity 11.4

1. *Statement of the problem:* The purpose of this experiment is to determine which of five label designs is preferred by customers across all of the company's five outlets, across the entire city (five boroughs) of New York. The target population consists of all sales in the five outlets and five boroughs. The criterion measure for this study is sales level for the product in the first week of exposure.
2. *Research hypothesis:* There is no significant difference in average sales for the first week of exposure for the five label designs.
3. *Statistical hypothesis:*

$$H_0: \mu_1 = \mu_2 = \mu_3 = \mu_4 = \mu_5$$
$$H_1: \mu_1 \neq \mu_2 \neq \mu_3 \neq \mu_4 \neq \mu_5$$

4. Seven-step Procedure (see Fig. E.2).

Case Study 4—Self-Review Activity 11.5

1. The brackets indicate that F and p values for the blocked effects should not be included in the report since we already know there is an impact (or we would not be blocking) and we are not really interested in the block effects.
2. The statistical analysis yields an F value of 1.108 with a p value of .397 which indicates that we should accept the null hypothesis, H_0. There is no significant difference between the five label designs in terms of customer preference.

Step Number	Description	Participant Response
Step I	State the null and research hypotheses.	$H_0:$ $\mu_1 = \mu_2 = \mu_3 = \mu_4 = \mu_5$ $H_1:$ $\mu_1 \neq \mu_2 \neq \mu_3 \neq \mu_4 \neq \mu_5$
Step II	State the maximum risk of committing a Type I error.	$\alpha = 0.10$
Step III	State the associated test statistic.	$F = MS_T / MS_E$
Step IV	Identify the random sampling distribution of the test statistic when H_0 is true.	$F \underset{d}{=} F(4, 12)$ df when H_0 is true
Step V	State the critical value for rejecting the null hypothesis.	Reject H_0 if $p < 0.10$
Step VI	Calculate the value of the test statistic from the sample data.	
Step VII	Analyze the results and make an appropriate inference related to the hypotheses tested.	$p =$ Discussion:

Figure E.2 Seven-step Procedure for Hypothesis Testing for Self-Review Activity 11.4.

Case Study 5—Self-Review Activity 11.6

1. See seven-step procedure in Fig. E.3.
2. To determine which combination of glue type and glue temperature maximize adhesion, a post-hoc analysis is required. The ANOVA presented in case study 5 showed that there was no interaction between glue type and temperature. Therefore, each factor can be assessed across the other factor. The means plot in Fig. E.4 illustrates the lack of interaction.

Step Number	Description	Participant Response
Step I	State the null and research hypotheses.	$H_0: I_{GT} = 0$ $H_1: I_{GT} \neq 0$
Step II	State the maximum risk of committing a Type I error.	$\alpha = 0.10$
Step III	State the associated test statistic.	$F = MS_I / MS_{Res}$
Step IV	Identify the random sampling distribution of the test statistic when H_0 is true.	$F \underset{=}{d} F(4, 1287)$ df when H_0 is true
Step V	State the critical value for rejecting the null hypothesis.	Reject H_0 if $p < 0.10$
Step VI	Calculate the value of the test statistic from the sample data.	$F = 1.645$
	Analyze the results and make an appropriate inference related to the hypotheses tested.	$p = 0.160$ Discussion: Do not reject H_0. There is no significant interaction between glue type and glue temperature (which means we can assess each factor across levels of the other factor).

Figure E.3 Seven-step Procedure for Hypothesis Testing for Self-Review Activity 11.6.

Step Number	Description	Participant Response
Step I	State the null and research hypotheses.	H_0: $\mu_{GC} = \mu_{GX1} = \mu_{GX2}$ H_1: $\mu_{GC} \neq \mu_{GX1} \neq \mu_{GX2}$
Step II	State the maximum risk of committing a Type I error.	$\alpha = 0.10$
Step III	State the associated test statistic.	$F = MS_{Glue} / MS_{Res}$
Step IV	Identify the random sampling distribution of the test statistic when H_0 is true.	$F \underset{=}{d} F\ (2,\ 1287)$ df when H_0 is true
Step V	State the critical value for rejecting the null hypothesis.	Reject H_0 if $p < 0.10$
Step VI	Calculate the value of the test statistic from the sample data.	$F = 321.350$
	Analyze the results and make an appropriate inference related to the hypotheses tested.	$p = .000$ Discussion: Reject H_0. We have sufficient statistical evidence to infer that there is a significant difference in adherence strength among the three glue types.

Figure E.3 (*Continued*)

Step Number	Description	Participant Response
Step I	State the null and research hypotheses.	$H_0: \mu_{TL} = \mu_{TT} = \mu_{TH}$
		$H_1: \mu_{TL} \neq \mu_{TT} \neq \mu_{TH}$
Step II	State the maximum risk of committing a Type I error.	$\alpha = 0.10$
Step III	State the associated test statistic.	$F = MS_{Temp} / MS_{Res}$
Step IV	Identify the random sampling distribution of the test statistic when H_0 is true.	$F \underset{=}{d} F(2, 1287)$ df when H_0 is true
Step V	State the critical value for rejecting the null hypothesis.	Reject H_0 if $p < 0.10$
Step VI	Calculate the value of the test statistic from the sample data.	$F = 22.671$
	Analyze the results and make an appropriate inference related to the hypotheses tested.	$p = .000$ Discussion: Reject H_0. We have sufficient statistical evidence to infer that there is a significant difference in adherence strength among the three glue temperatures.

Figure E.3 (*Continued*)

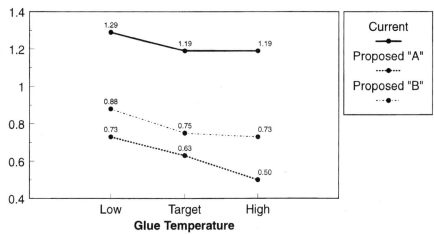

Figure E.4 Means Plot: Force by Glue Type and Temperature.

The ANOVA presented in case study 5 also indicated that there was a significant difference between the levels of glue type (across temperature) and between the levels of glue temperature (across glue type). Since each factor has three levels, a significant difference may mean that all three levels are significantly different or that one level is significantly different from the other two. To determine which levels differ significantly, a post-hoc analysis is required. The post-hoc procedure that follows must be used with caution so as not to confound the variability of the significant other factor in the error term of the post-hoc. The most conservative approach is the Scheffe procedure—method I which uses a penalty factor to significantly reduce the likelihood of committing a Type I error. This procedure can be performed manually as follows using $J - 1$ orthogonal contrasts (where J is the number of levels of the factor being tested).

The F statistic is calculated as

$$F = \frac{MS_\psi}{MS_R}$$

where MS_R is the mean square residual value used as the appropriate error term in testing the significance of the factor (.130 in this case from the ANOVA table in Fig. 11.28) and

$$MS_\psi = \frac{(\sum cwT)^2}{\sum cw^2} = \frac{n(\sum cw\overline{X})^2}{\sum cw^2}$$

where $\underline{T}$ = level total
$\underline{X}$ = level average
n = sample size used to calculate level average
cw = contrast weight

The F statistic is compared to an F critical value calculated as:

$$(J - 1)F_{cr(J-1, Jn-J)df}$$

which is $(J - 1)$ multiplied by the F critical value from the F table (App. C) with $J - 1$ and $J_n - J$ degrees of freedom at the appropriate significance level (in this case, $\alpha = .10$). Note that $J_n - J$ corresponds to the number of degrees of freedom associated with the MS_R term (1287 in this case, from the ANOVA table in Fig. 11.28).

In this case, to determine which levels of glue type are significantly different, the $J - 1 = 2$ orthogonal contrasts to be tested (based on the means plot in Fig. E.4) are (2 −1 −1) and (0 −1 1). The first contrast will compare glue type 1 (our current glue type) with types 2 and 3. The second contrast will compare glue types 2 and 3. The test would be performed as follows:

Contrast 1 (2 −1 −1):

$$MS_\psi = \frac{432[2(1.22) + (-1).62 + (-1).79]^2}{(2^2 + 1^2 + 1^2)} = 76.3848$$

Therefore,

$$F = \frac{MS_\psi}{MS_R} = \frac{76.3848}{.130} = 587.5754$$

This value is compared to the F critical value calculated as

$$(J - 1)\, F_{cr(J-1,\, Jn-J)df} = (3 - 1)\, F_{\alpha\, =\, .10\,\text{with}\,(3,\,1287)df} = 2 \times 2.60 = 5.2$$

Clearly, the calculated F statistic (587.5754) is greater than the F critical value (5.2), so there is sufficient statistical evidence to infer that level 1 (our current glue type) is significantly different from the other two glue types.

Contrast 2 (0 −1 1):

The next question is "Are glue types 2 and 3 significantly different?" This question is addressed with the second contrast (0 −1 1). Using the same formulas, we find the F statistic corresponding to that contrast is 282.4615 and is also greater than the F critical value.

Therefore, glue type is a significant (and important) factor affecting mean force with level 1 (our current glue type) giving the highest mean force of 1.22. The mean force corresponding to level 1 is significantly greater than the mean force for the other two glue types. Glue type at level 3 (GX2 = proposed B) has the second highest mean force at .79 which is significantly higher than the mean force of glue type at level 2 (GX1 = proposed A) at .62.

A similar analysis conducted on glue temperature indicates that all three levels of glue temperature yield significantly different mean force values with glue temperature 1 (low) giving the (significantly) highest mean force value at .97. Glue temperature 2 (target) yields a significantly higher mean force value (.86) than glue temperature 3 (high) which has a mean force value of .80.

This post-hoc analysis completes the study with respect to central tendency. To select "optimum set-points," an analysis of dispersion would be required. Such an analysis is beyond the scope of this book. At this point, if glue type and temperature were not significant with respect to variability, then the best selection for set up would be glue type 1 (our current glue) at temperature 1 (low).

Case Study 6—Self-Review Activity 11.7

1. *Statement of the problem:* The purpose of this experiment is to determine whether percent filler mixed in the polymer and/or cycle time affects the part quality of injected molded parts and to determine optimum set points. The target population for this study consists of all injected molded parts produced at the Lincoln plant with part quality identified as the dependent variable. The criterion measure is density (higher is better).
2. *Research hypotheses:*
   a. There is no significant difference in average density of injected molded parts made with two different levels of filler.
   b. There is no significant difference in average density of injected molded parts made with four different cycle times.
   c. There is no significant interaction between filler and cycle time for average density of injected molded parts.
   d. There is no significant difference in variability of density of injected molded parts made with two different levels of filler.
   e. There is no significant difference in variability of density of injected molded parts made with four different cycle times.
   f. There is no significant interaction between filler and cycle time for variability of density of injected molded parts.
3. *Statistical hypotheses:*
   a. H_0: $\mu_{F1} = \mu_{F2}$
      H_1: $\mu_{F1} \neq \mu_{F2}$
   b. H_0: $\mu_{C1} = \mu_{C2} = \mu_{C3} = \mu_{C4}$
      H_1: $\mu_{C1} \neq \mu_{C2} \neq \mu_{C3} \neq \mu_{C4}$
   c. H_0: $I_{FC} = 0$ (interaction of means)

H_1: $I_{FC} \neq 0$

d. H_0: $\sigma^2_{F1} = \sigma^2_{F2}$
   H_1: $\sigma^2_{F1} \neq \sigma^2_{F2}$
e. H_0: $\sigma^2_{C1} = \sigma^2_{C2} = \sigma^2_{C3} = \sigma^2_{C4}$
   H_1: $\sigma^2_{C1} \neq \sigma^2_{C2} \neq \sigma^2_{C3} \neq \sigma^2_{C4}$
f. H_0: $I_{FC} = 0$ (interaction of ln σ^2)
   H_1: $I_{FC} \neq 0$

where I_{FC} is the interaction between filler (percent filler in polymer) and cycle (cycle time).

4. The completed means plot is shown in Fig. E.5.
5. *Means analysis* (hypotheses *a* through *c* in item 3): Section 1 of the analysis in Fig. 11.31 shows a significant interaction between percent filler and cycle time with respect to part density. This means that we do not have enough information at this point to make conclusions about the optimum level of cycle time and percent filler to maximize average part density. The initial ANOVA (Fig. 11.31) identified a significant interaction between cycle time and % filler. The means plot in Fig. E.5 indicates that the interaction may be due to cycle time 1 (6). This post-hoc analysis would begin with a test to determine whether the interaction exists between filler and cycle for cycle = 2 (8), 3 (10), and 4(12). This post-hoc analysis would test the following hypotheses:

a H_0: $I_{FC} = 0$
   H_1: $I_{FC} \neq 0$ for cycle = 2, 3, and 4

If there is no significant interaction:

b. H_0: $\mu_{F1} = \mu_{F2}$
   H_1: $\mu_{F1} \neq \mu_{F2}$ for cycle = 2, 3, and 4

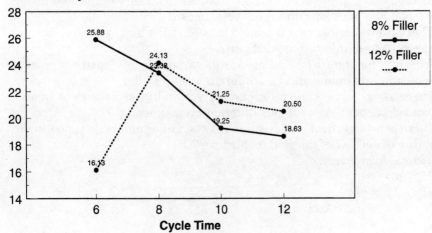

Figure E.5 Means Plot: Density by Percent Filler and Cycle Time.

c. H_0: $\mu_{C2} = \mu_{C3} = \mu_{C4}$
   H_1: $\mu_{C2} \neq \mu_{C3} \neq \mu_{C4}$
d. H_0: $\mu_{F1} = \mu_{F2}$
   H_1: $\mu_{F1} \neq \mu_{F2}$ for cycle = 1

The seven-step procedures for hypothesis testing for these tests (based on the statistical analysis that follows) are shown in Fig. E.6.

A Two-Way ANOVA of filler at both levels and cycle time at levels 2, 3, and 4 is shown in Fig. E.7. A p value of .910 indicates that there is no significant

Step Number	Description	Participant Response
Step I	State the null and research hypotheses.	H_0: $I_{FC} = 0$ H_1: $I_{FC} \neq 0$ for cycle = 2, 3, 4
Step II	State the maximum risk of committing a Type I error.	$\alpha = 0.10$
Step III	State the associated test statistic.	$F = MS_{FC} / MS_{Res}$
Step IV	Identify the random sampling distribution of the test statistic when H_0 is true.	$F \stackrel{d}{=} F(2, 42)$ df when H_0 is true
Step V	State the critical value for rejecting the null hypothesis.	Reject H_0 if $p < 0.10$
Step VI	Calculate the value of the test statistic from the sample data.	$F = 0.10$
Step VII	Analyze the results and make an appropriate inference related to the hypotheses tested.	$p = 0.910$ Discussion: Accept H_0. There is no significant interaction between filler and cycle for cycle = 2, 3, and 4.

Figure E.6 Seven-step Procedures for Post-hoc Analysis for Case Study 5.

Step Number	Description	Participant Response
Step I	State the null and research hypotheses.	$H_0: \mu_{F1} = \mu_{F2}$
		$H_1: \mu_{F1} \neq \mu_{F2}$ for cycle = 2, 3, 4
Step II	State the maximum risk of committing a Type I error.	$\alpha = 0.10$
Step III	State the associated test statistic.	$F = MS_C / MS_{Res}$
Step IV	Identify the random sampling distribution of the test statistic when H_0 is true.	$F \underset{d}{=} F(1, 42)$ df when H_0 is true
Step V	State the critical value for rejecting the null hypothesis.	Reject H_0 if $p < 0.10$
Step VI	Calculate the value of the test statistic from the sample data.	$F = 64.14$
Step VII	Analyze the results and make an appropriate inference related to the hypotheses tested.	$p = .000$ Discussion: Reject H_0. We have sufficient statistical evidence to infer that there is a significant difference in part density between filler 1 and filler 2 across cycles 2, 3, and 4.

Figure E.6 (*Continued*)

Step Number	Description	Participant Response
Step I	State the null and research hypotheses.	H_0: $\mu_{C2} = \mu_{C3} = \mu_{C4}$ H_1: $\mu_{C2} \neq \mu_{C3} \neq \mu_{C4}$
Step II	State the maximum risk of committing a Type I error.	$\alpha = 0.10$
Step III	State the associated test statistic.	$F = MS_C / MS_{Res}$
Step IV	Identify the random sampling distribution of the test statistic when H_0 is true.	$F \underset{d}{=} F(2, 42)$ df when H_0 is true
Step V	State the critical value for rejecting the null hypothesis.	Reject H_0 if $p < 0.10$
Step VI	Calculate the value of the test statistic from the sample data.	$F = 93.53$
Step VII	Analyze the results and make an appropriate inference related to the hypotheses tested.	$p = .000$ Discussion: Reject H_0. We have sufficient statistical evidence to infer that there is a significant difference in part density among cycles 2, 3, and 4.

Figure E.6 *(Continued)*

Step Number	Description	Participant Response
Step I	State the null and research hypotheses.	$H_0: \mu_{F1} = \mu_{F2}$ $H_1: \mu_{F1} \neq \mu_{F2}$ for cycle = 1
Step II	State the maximum risk of committing a Type I error.	$\alpha = 0.10$
Step III	State the associated test statistic.	$t = \dfrac{\overline{X}_{F1} - \overline{X}_{F2}}{\sqrt{\dfrac{S_P^2}{n-2}}}$
Step IV	Identify the random sampling distribution of the test statistic when H_0 is true.	$t \underset{=}{d} t(14)$ df when H_0 is true
Step V	State the critical value for rejecting the null hypothesis.	Reject H_0 if $p < 0.10$
Step VI	Calculate the value of the test statistic from the sample data.	$t = 30.43$
Step VII	Analyze the results and make an appropriate inference related to the hypotheses tested.	$p = .000$ Discussion: Reject H_0. We have sufficient statistical evidence to infer that there is a significant difference between filler 1 and filler 2 when cycle = 1.

Figure E.6 (*Continued*)

SPSS for MS WINDOWS Release 6.1

- - Description of Subpopulations - -

Summaries of   DENSITY
By levels of   CYCLE   number of cycles
                 FILLER   percentage of filler

Variable	Value	Label	Mean	Std Dev	Cases
For Entire Population			21.0156	3.0315	64
CYCLE	1.00	6	21.0000	5.0728	16
FILLER	1.00	8%	25.8750	.6409	8
FILLER	2.00	12%	16.1250	.6409	8
CYCLE	2.00	8	23.2500	1.2910	16
FILLER	1.00	8%	22.3750	.9161	8
FILLER	2.00	12%	24.1250	.9910	8
CYCLE	3.00	10	20.2500	1.2383	16
FILLER	1.00	8%	19.2500	.4629	8
FILLER	2.00	12%	21.2500	.8864	8
CYCLE	4.00	12	19.5625	1.2093	16
FILLER	1.00	8%	18.6250	.5175	8
FILLER	2.00	12%	20.5000	.9258	8

******Analysis of Variance -- design 1******

Tests of Significance for DENSITY using UNIQUE sums of squares

Source of Variation	SS	DF	MS	F	Sig of F
WITHIN+RESIDUAL	27.62	42	.66		
CYCLE	123.04	2	61.52	93.53	.000
FILLER	42.19	1	42.19	64.14	.000
CYCLE BY FILLER	.13	2	.06	.10	.910
(Model)	165.35	5	33.07	50.28	.000
(Total)	192.98	47	4.11		

R-Squared =   .857
Adjusted R-Squared = .840

Figure E.7 ANOVA for Filler (1,2) and Cycle (2,4).

Means analysis of filler (for cycle times 2, 3, and 4):

```
Summaries of      DENSITY
By levels of      FILLER    percentage of filler


Variable    Value Label              Mean      Std Dev   Cases

For Entire Population               21.0208    2.0263     48

FILLER      1.00  8%                20.0833    1.7917     24
FILLER      2.00  12%               21.9583    1.8292     24

Total Cases = 48
```

Figure E.7 (*Continued*)

interaction between filler and cycle time for cycle times 2, 3, and 4. Therefore, hypothesis 2 to compare levels of filler across cycle times 2, 3, and 4 can be tested. Since there are only two levels of filler and the ANOVA table shows that filler is significant, we can conclude that filler 2 (12 percent) gives significantly higher density values than filler 1 (8 percent) when cycle time is at level 2, 3, or 4. Since both factors are significant, the appropriate post-hoc analysis to determine which levels of cycle time are significantly different (hypothesis 3) would be the Scheffe procedure, method I that was presented in the last case study.

The next step (hypothesis 4) is to compare filler 1 versus filler 2 for cycle = 1. This analysis (Fig. E.8) shows that (with a p value = .000) there is a signif-

```
t-tests for Independent Samples of FILLER    percentage of filler


               Number
Variable       of Cases    Mean     SD      SE of Mean

DENSITY

8%               8        25.8750   .641      .227
12%              8        16.1250   .641      .227


        Mean Difference = 9.7500

        Levene's Test for Equality of Variances: F= .000  P= 1.000


        t-test for Equality of Means                 95%
        Variances    t-value   df   2-Tail Sig   SE of Diff   CI for Diff

        Equal         30.43    14     .000         .320       (9.063, 10.437)
        Unequal       30.43    14     .000         .320       (9.063, 10.437)
```

Figure E.8 t Test for Filler (1,2) within Cycle = 1.

icant difference between the two filler levels when cycle time = 1 (6). Filler 1 (8 percent) with an average density = 25.88 has a significantly higher part density than filler 2 (12 percent) which averages 16.13.

Since our original question was "Which combination of percent filler and cycle time will lead to higher part density?" the post-hoc analysis indicates that filler 1 (8 percent) and cycle time 1 (6) gives the (significantly) highest mean density value at 25.88.

Variance Analysis (hypotheses *d, e,* and *f* in item 3): From section 2 of the analysis presented in Fig. 11.31, we know that there is no interaction between filler and cycle with respect to the variability of part density. This means that each of the main effects can be assessed across levels of the other factor. The analysis shows that cycle time is not a significant factor but filler is significant with filler 1 (8 percent) leading to less variability in part density than filler 2 (12 percent). Therefore, in order to reduce variability, the optimum set point for this process is filler 1 = 8 percent.

The optimum set point to increase mean density values *and* decrease variability is filler at level 1 (8 percent) and cycle time at level 1 (6).

Case Study 7—Self-Review Activity 11.8

1. *Research hypotheses:*
    a. There is no significant difference in average moisture content between the five packaging types for soda crackers.
    b. There is no significant difference in average moisture content between boxes when package type 1 (control) is used, for all boxes in the population studied.
    c. There is no significant difference in average moisture content between boxes when package type 2 (wax paper) is used, for all boxes in the population studied.
    d. There is no significant difference in average moisture content between boxes when package type 3 (metal foil) is used, for all boxes in the population studied.
    e. There is no significant difference in average moisture content between boxes when package type 4 (plastic) is used, for all boxes in the population studied.
    f. There is no significant difference in average moisture content between boxes when package type 5 (metal foil and plastic) is used, for all boxes in the population studied.
2. The statistical analysis shows that there is a significant difference in moisture content box to box for each package type and there is a significant difference in moisture content between the five package types. Figure E.9 shows the means plot for moisture content by package type.

The means plot indicates that type 1 (control) seems to have a higher moisture content than the other types of packaging. Therefore, the appropriate

Mean Moisture Content

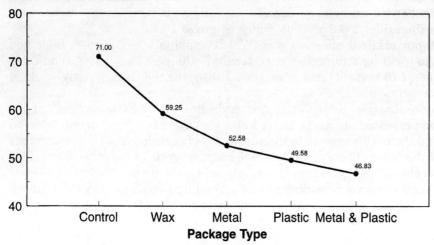

Figure E.9 Means Plot: Moisture by Package Type.

post-hoc analysis would begin by determining if there is a significant difference between packaging types 2, 3, 4, and 5. If not, test whether there is a significant difference between type 1 and types 2, 3, 4, and 5. Figure E.10 shows the data analysis for the first test.

The ANOVA table comparing packaging types 2, 3, 4, and 5 shows that (since $p = .136$, $p > .10$), we should accept the null hypothesis (that the packaging types are equal). We do not have sufficient statistical evidence to infer that there is a significant difference among package types 2 through 5. Our conclu-

SPSS for MS WINDOWS Release 6.1

* * * * * * A n a l y s i s   o f   V a r i a n c e -- design 1 * * * * * *

Tests of Significance for MOISTURE using UNIQUE sums of squares

Source of Variation	SS	DF	SS	F	Sig of F
WITHIN CELLS	278.00	32	8.69		
BOX	38.56	3	12.85	1.48	.239
PACKAGE BY BOX (ERROR 1)	1283.19	9	142.58	16.41	.000

Error 1	1283.19	9	142.58		
PACKAGE	1025.06	3	341.69	2.40	.136

Figure E.10 ANOVA for Package Types 2, 3, 4, and 5.

SPSS for MS WINDOWS Release 6.1

******Analysis of Variance -- design 1******

Tests of Significance for MOISTURE using UNIQUE sums of squares

Source of Variation	SS	DF	MS	F	Sig of F
WITHIN CELLS	142.00	16	8.88		
BOX	522.46	3	174.15	19.62	.000
PACKAGE BY BOX (ERROR 1)	51.79	3	17.26	1.95	.163

Error 1	51.79	3	17.26		
PACKAGE	828.37	1	828.37	47.98	.006

Figure E.11 ANOVA for Package Type 1 versus 2 through 5.

sion, then, is that the moisture content is not significantly different among packaging types 2, 3, 4, and 5. Therefore, the difference seen in the initial analysis (Fig. 11.35) must be due to the difference between packaging type 1 and the other four.

The data analysis presented in Fig. E.11 tests the second hypothesis of the post-hoc study (packaging type 1 versus the other four packaging types). In this case, $p = .006$, which confirms that there is a significant difference between packaging type 1 and the other four types. Packaging type 1 (control) leads to a significantly higher moisture level than the other four packaging types (wax, metal, plastic, or metal and plastic.) We do not have sufficient statistical evidence to show any significant difference between those four packaging types, so the best choice to minimize moisture content would be any of the other four packaging types, but not the control packaging type (cardboard box). If the means of these four packaging types are not significantly different, then the next step in the post-hoc analysis would be to determine which package types offer the least amount of variability box to box. This box to box variability is measured by σ_T^2, the between-component variance that was tested to determine if the box effect (a random effect) was significant. Since we rejected the null hypothesis that $\sigma_T^2 = 0$, we can calculate and compare the σ_T^2 values directly.

Using a one-way ANOVA from SPSS for each of the five packaging types, we find the ANOVA tables in Fig. E.12. From the information from the ANOVA table, the between-component variance can be calculated for each packaging type. Since quality is generally defined as the reduction of variability around a customer-defined target, a lower between-component variance is preferred.

The formula for the between-component variance is

$$\sigma_T^2 = \frac{MS_B - MS_W}{K'}$$

SPSS for MS WINDOWS Release 6.1

PACKAGE: 1.00   Control

----- O N E W A Y -----

Variable   MOISTURE   Moisture Content
By Variable   BOX   Box (Random Effect)

Analysis of Variance

Source	D.F.	Sum of Squares	Mean Squares	F Ratio	F Prob.
Between Groups	3	352.6667	117.5556	10.7684	.0035
Within Groups	8	87.3333	10.9167		
Total	11	440.0000			

PACKAGE: 2.00   Wax

----- O N E W A Y -----

Variable   MOISTURE   Moisture Content
By Variable   BOX   Box (Random Effect)

Analysis of Variance

Source	D.F.	Sum of Squares	Mean Squares	F Ratio	F Prob.
Between Groups	3	221.5833	73.8611	10.8089	.0035
Within Groups	8	54.6667	6.8333		
Total	11	276.2500			

Figure E.12   One-way ANOVA of Box (1, 4) for Each Package Type.

where MS_B = mean square between from the ANOVA table
MS_W = mean square within from the ANOVA table

$$K' = \frac{1}{J-1} \left[\left(\sum_{j=1}^{J} n_j \right) - \sum_{j=1}^{J} nj^2 \Big/ \sum_{j=1}^{J} n_j \right]$$

PACKAGE: 3.00 Metal

----- O N E W A Y -----

Variable   MOISTURE   Moisture Content
By Variable   BOX   Box (Random Effect)

Analysis of Variance

Source	D.F.	Sum of Squares	Mean Squares	F Ratio	F Prob.
Between Groups	3	168.2500	56.0833	5.1769	.0280
Within Groups	8	86.6667	10.8333		
Total	11	254.9167			

PACKAGE: 4.00 Plastic

----- O N E W A Y -----

Variable   MOISTURE   Moisture Content
By Variable   BOX   Box (Random Effect)

Analysis of Variance

Source	D.F.	Sum of Squares	Mean Squares	F Ratio	F Prob.
Between Groups	3	445.5833	148.5278	10.1269	.0042
Within Groups	8	117.3333	14.6667		
Total	11	562.9167			

Figure E.12 (*Continued*)

Since the sample sizes and number of levels (of box) are the same for each packaging type, K' will be the same for each.

$$K' = \frac{1}{4-1}\left[(3+3+3+3) - \frac{(3^2+3^2+3^2+3^2)}{(3+3+3+3)}\right] = 3$$

For package type 1,

$$\sigma_T^2 = \frac{117.5556 - 10.9167}{3} = 35.5463$$

Likewise, $\sigma_T^2 = 22.3426$ for package type 2, $\sigma_T^2 = 15.0833$ for package type 3, $\sigma_T^2 = 44.6204$ for package type 4, and $\sigma_T^2 = 53.2315$ for package type 5. Advanced formulas are available to calculate confidence intervals for the between-component variances and to perform statistical tests to determine if they are significantly different. If there is a significant difference, we would select the package type that has the lowest between-component variance, which in this case is package type 3 (metal).

Case Study 8—Self-Review Activity 11.9

1. The F values are compared to the F critical values found in App. C for $\alpha = 0.10$. $F(1, 16) = 3.05$ and $F(3, 16) = 2.49$. The data indicate that factors D, E, and F and the AB, GI, and GH interactions are significant. [Factors A, H, and I also have F values greater than the F critical values, but since the AB, GI, and GH interaction effects are significant, the main effects (A, B, G, H, and I) cannot be assessed across the other factors involved in a significant interaction.] Screening experiments are used to determine which factors are not only significant, but also important. Importance is measured by rho %, which is a measure of the percent of variation explained by that factor. Factors (and interactions) with rho % > 10 percent are considered important if we are running the experiment to determine special causes of variation (which are few in number but great in impact). Factors (and interactions) with rho > 5 percent are considered important if we are running the experiment to determine common causes of variation (which are large in number but relatively low in impact). In this case, we are assessing a process that is in a state of control to determine which factors affect the process in an effort to set optimum set points. So, on the basis of the rho % values, the factors and interactions that are both significant and important are D, E, and AB, where:

A = foil paper
B = cup burrs
D = lacquer vendor
E = thickness

The means plots show that, since a higher seating force is preferred, foil paper present (A = 1), cup burr level 2 (B = 2), lacquer vendor A and/or B (D = 1 and/or 2), and thickness level 2 (E = 2) are tentative optimum settings. In the case of lacquer vendor (since there are more than two levels), a post-hoc analysis would be required to determine whether there is a significant difference between vendors A and B. These settings are called *tentative* because a confirmation experiment is necessary to confirm the results.

2. It is not appropriate to make a process change until a confirmation experiment is completed.

3. The confirmation experiment should test the following factors and levels:

A = foil paper at 2 levels
B = cup burrs at 2 levels
D = lacquer vendor at 4 levels
E = thickness at 2 levels

A full factorial design would require $2 \times 2 \times 4 \times 2 = 32$ runs. If this many runs is not economically feasible, it may be possible to run a one-half fractional design with 16 runs to test all main effects and all two-way interactions. However, note that a confirmation experiment is always required after a fractional design, so this approach may result in more runs over the course of both experiments than if the full factorial is used. Industrial research constantly balances the cost of acquiring data with the value of the information received from the analysis. If a full factorial design is used (all 32 runs), the ANOVA table would appear as follows:

```
*** ANALYSIS OF VARIANCE ***
        FORCE
BY   A    FOIL PAPER
     B    CUP BURRS
     D    LACQUER VENDOR
     E    THICKNESS
```

Source of Variation	Sum of Squares	DF	Mean Square	F	Sig of F
Main Effects					
A					
B					
D					
E					
2-way Interactions					
AB					
AD					
AE					
BD					
BE					
DE					
3-way Interactions					
ABD					
ABE					
BDE					
4-way Interactions					
ABDE					
Residual					
Total					

Bibliography

Ackoff, R., 1953. *The Design of Social Research.* Chicago: University of Chicago Press.
Anderson, V., and McLean, R., 1974. *Design of Experiments: A Realistic Approach.* New York: Marcel Dekker.
Bainbridge, T., 1965. "Staggered, nested designs for estimating variance components," *Industrial Quality Control,* Vol. 22 (No. 1): pp. 12–20.
Box, G., Hunter, W., and Hunter, J., 1978. *Statistics for Experimenters.* New York: John Wiley and Sons.
Bratcher, T. L., Moran, M. A., and Zimmer, W. J., 1970. "Tables of sample sizes in the Analysis of Variance," *Journal of Quality Technology,* Vol. 2 (No. 3): pp. 156–164.
Campbell, D., and Stanley, J., 1963. *Experimental and Quasi-Experimental Designs for Research.* Chicago: Rand McNally College Publishing Company.
Dowdy, S., and Weardon, S., 1983. *Statistics for Research.* New York: John Wiley and Sons.
Duncan, W., and Luftig & Warren International. 1995. *Total Quality: Key Terms and Concepts.* New York: American Management Association.
Gibbons, J., 1976. *Nonparametric Methods for Quantitative Analysis.* New York: Holt, Rinehart, and Winston.
Guilford, J. P., 1974. *Psychometric Methods.* New York: McGraw-Hill.
Hicks, C. R., 1964. *Fundamental Concepts in the Design of Experiments.* New York: Holt, Rinehart, and Winston.
Juran, J., Gryna, F., and Bingham, R., 1979. *Quality Control Handbook,* 3d ed. New York: McGraw-Hill.
Kirk, R., 1968. *Experimental Design: Procedures for the Behavioral Sciences.* Belmont, Calif.: Brooks/Cole Publishing.
Kotz, S., and Johnson, N., 1985. *Encyclopedia of Statistical Sciences,* Vol. 5. New York: John Wiley & Sons.
Luftig, J., 1989. *A Quality Improvement Strategy for Critical Product and Process Characteristics.* Southfield, Mich.: Luftig & Warren International.
Luftig, J., 1991. *Experimental Design and Industrial Statistics—Level II.* Southfield, Mich.: Luftig & Warren International.
Luftig, J., 1991. *Experimental Design and Industrial Statistics—Level III.* Southfield, Mich.: Luftig & Warren International.
Luftig, J., 1991. *Guidelines for a Practical Approach to Gauge Capability Analysis.* Southfield, Mich.: Luftig & Warren International.
Luftig, J., 1992. *Guidelines for a Practical Approach to the Assessment of Discrete Data Measurement Systems,* Vol. 1. Southfield, Mich.: Luftig & Warren International.
Mander, A. E., 1947. *Logic for the Millions.* New York: Philosophical Library.
Miller, D., 1964. *Handbook of Research Design and Social Measurement.* New York: David McKay.
Natrella, M., 1963. *Experimental Statistics.* National Bureau of Standards Handbook 91. Washington: Government Printing Office.
Reliability Office, NAAO, 1979. *Experimental Design.* Dearborn, Mich.: Ford Motor Company.
Rice, M., 1913. *Scientific Management in Education.* New York: Hinds, Noble, & Eldredge.
Sax, G., 1979. *Foundations of Educational Research.* Englewood Cliffs, N.J.: Prentice-Hall.
Siegel, S., and Castellan, N. J., 1988. *Nonparametric Statistics for the Behavioral Sciences,* 2d ed. New York: McGraw-Hill.
Spector, P., 1982. "Research Designs," Sage University paper. Beverly Hills: Sage Publications.
Stouffer and Bartky, *Chicago Tribune,* November 2, 1940.
Wallis, W., and Roberts, H., 1962. *Statistics: A New Approach.* New York: Glencoe, Inc.
Wildt, A., and Ahtola, O., 1978. *Analysis of Covariance.* Beverly Hills, Calif.: Sage Publications.

Index

Accident sampling plans, 132–133
Action (data) sweep analysis, 26
Agreement research, 26–27, 132
Alias structure, in fractional factorials, 255, 257
Analysis of covariance (ANCOVA), 142, 264–265
Analytical research, 25–26
ANCOVA (*see* analysis of covariance (ANCOVA))
Anderson-Darling normality test, 194–195, 196
ANOVA (*see* one-way analysis of variance)
Assumption approach to independent variables, 116
Authoritative information in research, 9–10

Balanced incomplete block design, 102
Bartlett-Box F tests, 191
Block design, 40, 42, 92, 201–209
   multiple blocked effect, 215–221
Blocked factorial design, 103
Box-and-whisker normality test, 48, 50, 196
Burn-in tests, 91–92

Capability, in experimental design, 45
Case studies, 28, 63–64
Case studies in industrial experiments, 189–266
Causal vs. Coincidental relationships, 21–22, 28–32
Chance and random effects, 92
Classical designs, 120
Cluster random sampling, 134
Combination, treatments, 41
Common sense vs. Scientific method, 9
Communication skills, 13

Concordance analysis, 27
Concurrent correlation research, 28–29
Concurrent validity, 144–145
Confidence effect on outcomes, 96
Confounding effect, 44, 70, 77–87, 92–95
Consensus analysis, 26–27
Construct validity, 145
Content validity, 144
Contingency plan, 29, 30
Control data, 142
Controls/control groups, 44, 68–69, 70
Correlating (forming relationships) data, 13
Correlation coefficient, 41, 86
Covariance (*see* analysis of covariance (ANCOVA))
Covariates (independent variables), 108
Criterion data, 141
Criterion measure, 40, 105, 106–107, 108
Critical values for null hypotheses, 168
Crossed analytical models, 111–112
Cross-sectional studies, 28

Data collection, 8–9, 26, 141–142
   classifying data, 18
   control data, 142
   criterion data, 141
   descriptive data, 141
Data mining, 26
Deductions or hypotheses, 13
Deductive reasoning, 10–13, 35
Degrees of freedom, 53, 168
Delimiting research problems, 18–19
Delineating the problem, 18
Dependent (response) variables, 40, 62, 76, 105–106, 106–107, 108
Descriptive data, 141
Descriptive research, 27–28
Descriptive statistics, 83–84

Index

Deviation, 191
Disproportionate random sampling, 134
Distributions, 50, 51, 53, 56, 116–118

Efficiency index (EI) of design, 122
Empiricism, 13
Engineering logs, 156–159
Errors, 43–44
   (*See also* experimental design; validity checking)
Evolutionary operation (EVOP) technique, 265–266
Ex post facto studies, 31
Executing the experiment, 155–159
   complexity of experiment, 155–156
   engineering logs, 156–159
   responsibility, assigning responsibility, 155–156
Experimental (test) unit, 42
Experimental area, 40
Experimental design, 7, 12–13, 32, 39–45
   adequacy of chosen design, 122–123
   analysis of covariance (ANCOVA), 264–265
   approval of experiment, 99
   assessing the chosen design, 122–123
   assumption approach to independent variables, 116
   balanced incomplete block design, 102
   block design, multiple blocked effect, 215–221
   blocked factorial design, 103
   blocking independent variables, 115
   blocks, 40, 42, 92
   burn-in tests, 91–92
   case studies, 63–64
   causal vs. coincidental relationships, 21–22, 28–32
   chance and random effects, 92
   classical designs, 120
   combination, treatments, 41
   confidence effect on outcomes, 96
   confounding, 44, 70, 77–87, 92–95
   controls, 44, 68–69, 70
   correlation coefficient, 41, 86
   criterion measure, 40
   crossed analytical models, 111–112
   definition of terms in experiments, 39–40
   dependent (response) variables, 40, 76
   descriptive statistics, 83–84
   efficiency index (EI) of design, 122
   efficiency of chosen design, 122–123
   environment for experiment, 99
   error, experimental error, 43–44, 92, 99

Experimental design (*Cont.*):
   evolutionary operation (EVOP) technique, 265, 266
   executing the experiment, 155–159
   experimental (test) unit, 42
   experimental area, 40
   experimental vs. nonexperimental research, 20–25, 68
   external validity, 45
   external validity threats, 97–98
   extraneous, manipulable variable confounds denominator, 77–87
   extraneous, manipulable variable confounds numerator, 77
   factorial experiments, 41, 103
      fully crossed Type I, 221–229
      fully crossed type I, 229–240
      nested model III, 240–252
   fixed vs. random effect, 117–118
   fractional factorial design, 103, 120, 221, 253–263
   full-strength of treatment to test, 91–92
   Graeco-Latin square design, 221
   hierarchical design, arranging variables, 112–114
   history vs. outcomes, 71
   hyper-Graeco-Latin square design, 221
   hypotheses, 32-35
   independent variables, 41
   industrial experiments (*see* industrial experiment design)
   inference space, 43
   instrumentation, 72
   interactions affecting outcomes, 42, 70, 71, 97
   internal validity, 44–45
   internal validity, threats to, 69–74
   interpreting results, 99
   known, nonmanipulable nuisance variables, 116
   Latin square design, 102, 120, 215–221, 258
   levels, 41
   levels for treatment variables, 118–120
   life tests, 91–92
   limitation/control of independent variables, 114–115
   manipulating the outcome of experiments, 23–25, 61
   maturation, 72
   Monte Carlo simulation, 89–90
   mortality, 70, 73
   multiple treatment interference, 97–98
   nested design, 103, 112–114, 240–252

Experimental design (Cont.):
  nonparametric testing and analysis, 163, 165, 166
  notation in experimental design, 44
  number of experiments, 99
  one-group pretest-posttest design, 64–68, 70
  one-shot case studies, 63–64
  orthogonal array designs, 120, 258–261
  outcomes, 62
  outline of steps in, 100
  parameters, 44
  parametric testing and analysis, 162–163, 165
  partially balanced incomplete block design, 102
  Plackett-Burman extreme screening design, 120, 256–257
  planned grouping, 42–43, 92, 96
  planning checklist (see planning checklists)
  planning the experiment, 61–103
  plots, 40
  populations, 43
  posttest-only control group design, 70, 74–75, 99
  power effect on outcomes, 95–96
  pre-experimental design, 63, 68
  pretest-posttest control group design, 64–68, 70, 73–74, 98
  problem definition, 99
  process capability, 45
  pseudoexperimental design, 68
  qualitative vs. quantitative variables, 117
  quasi-experimental design, 63, 68
  random effects, 92, 240–252
  randomization, 42, 70, 85
  randomized block design, 102, 120
    matched pairs, 201–209
    multilevel, 210–215
  randomized design, 102, 190–200
  reactive arrangements, 97
  regression, statistical, 73
  repeated measures, 92
  replication of results, 43, 96
  reporting results, 171–172
  representative samples, 43
  research populations, 43
  run-in tests, 91–92
  sample size, 88, 91, 136–138
  sampling, 43, 131–139
  screening designs, 256–257
  selection bias, 70, 71
  sensitivity (power) of design, 77, 87–92, 95–96, 136–138

Experimental design (Cont.):
  sequence of experiments, 99
  significance levels, 86
  single-factor experiments, 41
  Solomon four-group design, 70, 75, 98
  Spearman Rank Correlations, 86
  split-plot design, 263
  split-split-plot designs, 263
  standard operating procedures (SOP), 45
  standardize-do-check-act checklist, 173
  static-group comparison design, 68–69
  statistical inference, 44
  statistically in or out of control, 45
  statistics in experiments, 39, 44
    (see also statistical analysis)
  success or failure of experiment, 99
  systematic error, 44
  target populations, 43
  testing, 72
  tools for sound experimentation, 42, 96
  traditional vs. modern methods, 99
  treatment combination, 41
  treatments, 40
  Type I errors, 43, 62–69
  Type II errors, 43, 76–95
  unexplained variation, 85
  universe (of experiment), 43
  validity, 44–45
    (see also validity checking)
  variables, 40, 41, 96
  variance effects, 79–83
  yields, 40
  Youden square design, 102, 120, 221
Experimental error, 92
Experimental research, 32
Experimental vs. Nonexperimental research, 20–25, 21, 68
External validity, 45
External validity threats, 97–98

Face validity, 145
Factorial experiments, 41, 103
  fully crossed Type I, 221–229
  fully crossed type I, 229–240
  nested model III, 240–252
Fixed vs. Random effect, 117–118
Focus groups, 132
Fractional factorial experiments, 103, 120, 221, 253–263
  alias structure, 255, 257
  fractions, 253
  higher-order interactions, 255
  higher-resolution designs, 257–263

Fractional factorial experiments (*Cont.*):
  Latin square design, 258
  one-eighth replicate, 254
  one-fourth replicate, 254
  one-half replicate, 253
  orthogonal array designs, 258–261
  Plackett-Burman extreme screening design, 256–257
  resolution, 255
  screening designs, 256–257
Full-strength of treatment to test, 91–92

Graeco-Latin square design, 221
Grouping, planned grouping, 42–43

Hierarchical design, arranging variables, 112–114
Higher-order interactions, in fractional factorials, 255
Histograms, 48, 49
Historical research, 25–26
Hyper-Graeco-Latin square design, 221
Hypotheses formulation, 13, 32–35, 47–48, 168
  clarity necessary, 34
  consistency of hypotheses with fact, 35
  deductive reasoning, 35
  null hypotheses, 167
  research hypotheses, 167
  testing hypotheses, 163, 167–169

Independent variables, 41, 105–106, 107–118
  (*See also* industrial experiment design; variables)
Inductive reasoning, 11–13
Industrial experiment design, 105–130
  adequacy of chosen design, 122–123
  analysis of covariance (ANCOVA), 264–265
  arranging variables in experiment, 109–116
  assessing the chosen design, 122–123
  assumption approach to independent variables, 116
  block design, multiple blocked effect, 215–221
  blocking variables, 115
  case studies, 189–266
  classical designs, 120
  classifying treatment variables, 116–118
  covariates (independent variables), 108
  criterion measures, 105, 106–107, 108

Industrial experiment design (*Cont.*):
  crossed analytical models, 111–112
  dependent variables, 105–107, 108
  design selection process, 120–121
  distributions, 116–118
  efficiency index (EI) of design, 122
  efficiency of chosen design, 122–123
  evolutionary operation (EVOP) technique, 265–266
  executing the experiment, 155–159
  extraneous effects, 116
  factorial experiments
    fully crossed Type I, 221–229, 229–240
    nested model III, 240–252, 240
  fixed vs. random effect, 117–118
  fractional factorial experiments, 120, 221, 253–263
  Graeco-Latin square design, 221
  hierarchical design, arranging variables, 112–114
  hyper-Graeco-Latin square design, 221
  incorporating independent variables in experiment, 110–112
  independent variable, 105–118
  known, nonmanipulable nuisance variables, 116
  Latin square design, 120, 215–221, 258
  levels for treatment variables, 118–120
  limitation/control of independent variables, 114–115
  nesting independent variables, 112–114, 240–252
  nuisance variables (independent variables), 108
  orthogonal array designs, 120, 258–261
  Plackett-Burman extreme screening design, 120, 256–257
  planning checklist, 124–130, 145, 147–154, 173–187
  problem definition, 105
  qualitative vs. quantitative variables, 117
  random effects, 240–252
  random sampling, 106
  randomized block design, 120
    matched pairs, 201–209
    multilevel, 210–215
  randomized design, 190–200
  reporting results, 171–172
  sampling, 106, 131–139
  screening designs, 256–257
  split-plot design, 263
  split-split-plot designs, 263
  standardize-do-check-act checklist, 173
  steps in design process, 105–106

Industrial experiment design (*Cont.*):
  treatment variables (independent variables), 108
  Youden square design, 120, 221
Inference space, 43
Instrumentation and validity, 72, 142–145
  concurrent validity, 144–145
  construct validity, 145
  content validity, 144
  equivalence, 144
  face validity, 145
  internal consistency or homogeneity, 144
  predictive validity, 144
  reliability, 143–144
  stability, 143
Interactions affecting outcomes, 42, 70, 71, 97
Internal validity, 44–45
  threats to, 69–74
Interval estimation, 47
Interval measurement, 162
Intuition in scientific research, 10

Judgment sampling plans, 132

Kurtosis, 192, 194–195

Latin square design, 102, 120, 215–221, 258
Levels for treatment variables, 41, 118–120
Levene test, 191, 193, 197
Life tests, 91–92
Limitation/control of independent variables, 114–115
Lin-Mudholkar normality test, 194–195, 196
Logbooks, engineering logs, 156–159
Logic, 10–13
Longitudinal studies, 28

Manipulating the outcome of experiments, 23–24, 61
Maturation, 72
Mean (average) differences, 47, 49–51
Mean deviation, 198–199
Mean testing, 194
Means analysis, 229–234
Measurement techniques, 15, 142–145, 161, 162
  criterion measures, 40, 105, 106–107, 108
  interval measurement, 162
  measurement recording forms, 146

Measurement techniques (*Cont.*):
  nominal measurement, 162
  ordinal measurement, 162
  ratio measurement, 162
Moments tests, 194–195
Monte Carlo simulation, 89–90
Mortality, 70, 73
Multiple treatment interference, 97–98
Multistage random sampling, 134

Natural work groups (NWG), 14–15
Nested designs, 103, 112–114, 240–252
Nominal measurement, 162
Nonparametric testing and analysis, 163, 165, 166
Nonprobalistic sampling, 131–133
Normal distributions, 47, 51
Normality, 191, 194–195
Notation in experimental design, 44
Nuisance variables (independent variables), 108
Null hypotheses, 51–52, 55, 167, 168–169

Objectivity of experimenters, 13
One-group pretest-posttest design, 64–68, 70
One-shot case studies, 63–64
One-way analysis of variance (ANOVA), 56, 57–59, 191, 193, 195, 197, 203–209, 211–215, 220–221, 224–228, 234–240, 263
Optimum allocation random sampling, 134
Ordinal measurement, 162
Orthogonal array designs, 120, 258–261

Parameters, 44, 162–163, 165, 166
Parametric testing and analysis, 162–163, 165
Parsimony, law or principle of, 13, 35
Partially balanced incomplete block design, 102
Philosophical research, 25–26, 132
Plackett-Burman extreme screening design, 120, 256–257
Plan-do-study-act (PDSA) management strategy, 8
Planned grouping, 42–43, 92, 96
Planning checklists, 8, 36–37, 124–130, 145, 147–154, 173–187
Plots, 40
Point estimation, 47
Populations, 43, 49, 51–53

Post-hoc analysis, 198–199, 211–215
Posttest-only control group design, 70, 74–75, 99
Power effect on outcomes, 95–96
  (*See also* sensitivity of design)
Predictive correlation research, 29–31
Predictive validity, 144
Pre-experimental design, 63, 68
Pretest-posttest control group design, 64–68, 70, 73–74, 98
Probabilistic sampling plans, 133–135
Probability (p value), 54
Problem solving through research, 14–20
  classifying data, 18
  delimiting research problems, 18–19
  delineating the problem, 18
  sample research problem statement, 19–20
  scientific method solutions, 17
  statement of problem, 17–20, 105
  trial-and-error solutions, 16–17
Process capability, 45
Proportionate random sampling, 134
Pseudoexperimental design, 68
Pseudoexperimental studies, 31
Purposive sampling plans, 132

Qualitative vs. Quantitative variables, 117
Quality improvement research, 14–15
Quality improvement strategy (QIS), 14–15
Quality improvement teams (QIT), 14–15
Quasi-experimental design, 31, 63, 68
Quota sampling plans, 132

Random effects, 92, 240–242
Random sampling distribution (RSD), 47, 50–56, 168
Random sampling, 49–50, 51, 133–136
Random vs. Fixed effect, 117–118
Randomization, 42, 70, 85, 102
Randomized block design, 102, 120
  matched pairs, 201–209
  multilevel, 210–215
Randomized design, 102, 190–200
Ratio measurement, 162
Reactive arrangements, 97
Reasoning, 10–13
Regression, statistical, 73
Relational research, 28–32
Reliability of instrumentation, 143–144
Repeated measures, 92
Replication of results, 43, 96
Reporting results, 171–172

Representative samples, 43
Research design process, 7–9
Research hypotheses, 167
Research populations, 43
Research question formulation, 32–35
Research study, 7–37
  action (data) sweep analysis, 26
  agreement research, 26–27
  analytical research, 25–26
  authoritative information, 9–10
  case studies, 28
  causal vs. coincidental relationships, 21–22, 28–32
  clarity of questions and hypotheses, 34
  common sense approach vs. scientific method, 9
  communication skills, 13
  concordance analysis, 27
  concurrent correlation research, 28–29
  consensus analysis, 26–27
  consistency of hypotheses with fact, 35
  correlating (forming relationships) data, 13
  cross-sectional studies, 28
  data classification, 18
  data collection, 18, 26
  data mining, 26
  deductions or hypotheses, 13
  deductive reasoning, 10–13, 35
  defects as subject for, 15
  delimiting research problems, 18–19
  delineating research problems, 18
  descriptive research, 27–28
  empiricism, 13
  experimental design in research, 7–9
  experimental vs. nonexperimental research, 20–25, 32
  experimentation, 12–13
  historical research, 25–26
  hypotheses development, 32–35
  inductive reasoning, 11–13
  intuition in scientific research, 10
  logic and reasoning, 10–13
  longitudinal studies, 28
  manipulating the outcome of experiments, 23–25, 61
  measurment techniques, 15
  objectivity of experimenters, 13
  observation, 12–13
  parsimony, law or principle of, 13, 35
  philosophical research, 25–26
  planning checklist, 36–37
  predictive correlation research, 29–31
  problem solving through, 14–20, 16

Research study (*Cont.*):
  problem statement for, 17–20
  product/process shortcomings as subject for, 14
  quality improvement efforts as subject for, 14–15
  quasi-experimental, ex post facto, or pseudoexperimental studies, 31
  questions, research questions, 32–35
  relational research, 28–32
  reporting results, 171–172
  sample research problem statement, 19–20
  scientific method of research, 1–14
  specificity of question/hypotheses, 35
  standardize-do-check-act checklist, 173
  status studies, 27–28
  theoretical framework, 20–32
  trial–and-error experiments, 11–12
  what it is, 7–20
Resolution, in fractional factorials, 255
Run-in tests, 91–92

Sampling, 43, 49–50, 131–139
  accident sampling plans, 132–133
  cluster random sampling, 134
  disproportionate random sampling, 134
  judgment sampling plans, 132
  multistage random sampling, 134
  nonprobalistic sampling, 131–133
  optimum allocation random sampling, 134
  probabilistic sampling plans, 133–135
  proportionate random sampling, 134
  purposive sampling plans, 132
  quota sampling plans, 132
  random sampling, 133–136
  simple random sampling, 134
  size of sample, 88, 91, 136–138
  stratified cluster random sampling, 134
  stratified random sampling, 134
  systematic random sampling, 134
  Type I and Type II errors associated with, 137–138
Sampling error (SE), 50, 54
Scientific method, 1–14, 32
  authoritative information in research, 9–10
  causal vs. coincidental relationships, 21–22, 28–32
  common sense approach to experiences, 9
  correlating (forming relationships) data, 13
  deductions or hypotheses, 13
  empiricism, 13
  experimentation in, 12–13

Scientific method (*Cont.*):
  hypotheses, 32–35, 47–48
  intuition vs., 10
  logic and reasoning, 10–13
  objectivity of experimenters, 13
  observation in, 12–13
  parsimony, law or principle of, 13, 35
  revelation vs., 10
  trial-and-error experiments, 11–12
Screening designs, 256–257
Selection bias, 70, 71
Sensitivity (power) of design, 87–92, 95–96, 136–138
Significance levels, 86
Simple random sampling, 134
Single-factor experiments, 41, 55–56
Skewness, 191, 194–195
Solomon four-group design, 70, 75, 98
Spearman Rank Correlations, 86
Specificity of question/hypotheses, 35
Split-plot design, 263
Split-split-plot designs, 263
Standard deviation, 49, 50
Standard operating procedures (SOP), 45
Standard statistical software packages (SSSP), 48
Standardize-do-check-act checklist, 173
Static-group comparison design, 68–69
Statistical analysis, 39, 44, 47–59, 161–169
  analysis of covariance (ANCOVA), 142, 264–265
  Anderson-Darling normality test, 194–195, 196
  Bartlett-Box F tests, 191
  box-and-whisker normality test, 48, 50, 196
  critical values for null hypotheses, 168
  degrees of freedom, 53, 168
  descriptive statistics, 83–84
  deviation, 191
  distributions, 50, 51, 53, 56
  evolutionary operation (EVOP) technique, 265–266
  histograms, 48, 49
  hypotheses for statistical analysis, 47–48
  hypothesis testing, 163, 167–169
  in or out of control, 45
  interval estimation, 47
  interval measurement, 162
  kurtosis, 192, 194–195
  Levene test, 191, 193, 197
  Lin-Mudholkar normality test, 194–195, 196
  mean (average) difference, 47, 49–51

Statistical analysis (*Cont.*):
  mean deviation, 198–199
  mean testing, 194
  means analysis, 229–234
  measurement scales, 161, 162
  moments tests, 194–195
  nesting, 240–252
  nominal measurement, 162
  nonparametric testing and analysis, 163, 165, 166
  normal distributions, 47, 51
  normality, 191, 194–195
  null hypotheses, 51–52, 55, 167, 168–169
  one-way analysis of variance (ANOVA), 56, 57–59, 191, 193, 195, 197, 203–209, 211–215, 220–221, 224–228, 234–240, 263
  ordinal measurement, 162
  parametric testing and analysis, 162–163, 165
  point estimation, 47
  population differences, 51–53
  populations, 49
  post-hoc analysis, 198–199, 211–215
  probability (p value), 54
  problem statement for statistical analysis, 47–48
  random effects, 240–252
  random samples, 49–51
  random sampling distribution (RSD), 47, 50–56, 168
  ratio measurement, 162
  regression, 73
  research hypotheses, 167
  sampling (*see* sampling)
  sampling error (SE), 50, 54
  selecting proper testing method, 161, 164
  single-factor experiments, 55–56
  skewness, 191, 194–195
  standard deviation, 49, 50
  standard statistical software packages (SSSP), 48
  Student-Newman-Keuls mean deviation procedure, 198–199, 200
  t distributions/testing, 47, 53–54, 56, 168, 203–209
  Type I error risks, 167
  validity, 55
  variance, 191, 194, 197–198, 234–240
Statistical inference, 44
Statistical regression, 73
Statistically in or out of control, 45
Status studies, 27–28
Stratified cluster random sampling, 134
Stratified random sampling, 134
Student-Newman-Keuls mean deviation procedure, 198–199, 200
Systematic error, 44
Systematic random sampling, 134

T distributions/testing, 47, 53–54, 56, 168
  matched pairs, 203–209
Taguchi arrays (*see* orthogonal arrays)
Target populations, 43
Testing, 72
Treatment combination, 41
Treatment variables (independent variables), 108
Treatments, 40
Trial-and-error experiments, 11–12
Type I errors, 43, 62–69, 167
Type II errors, 43, 76–95

Unexplained variation, 85
Universe (of experiment), 43

Validity/validity checking, 44–45, 55, 141–154
  (*see also* experimental design)
  chance and random effects, 92
  concurrent validity, 144–145
  confidence effect on outcomes, 96
  confounding, 92–95
  construct validity, 145
  content validity, 144
  correlation coefficients, 86
  data validity, 141–142
  experimental error, 92
  external validity threats, 97–98
  extraneous, manipulable variable confounds denominator, 77–87
  face validity, 145
  history vs. outcomes, 71
  instrumentation, 72, 142–145
  interactions affecting outcomes, 71, 97
  internal validity, threats to, 69–74
  maturation, 72
  measurement recording forms, 146
  measurement techniques, 142–145
  mortality, 73
  multiple treatment interference, 97–98
  planning checklist, 145, 147–154
  posttest-only control group design, 74–75
  power effect on outcomes, 95–96
  predictive validity, 144

Validity/validity checking (*Cont.*):
  pretest-posttest control group design, 73–74
  protecting against threats to internal validity, 73–76
  random effects, 92
  randomization, 85
  reactive arrangements, 97
  regression, statistical, 73
  reliability of instrumentation, 143–144
  repeated measures, 92
  sample size, 88, 91
  selection bias, 71
  sensitivity (power) of design, 87–92, 95–96
  significance levels, 86
  Solomon four-group design, 75
  testing, 72
  Type II errors, 76–95
  unexplained variation, 85
  variance effects, 79–83
Variables, 40, 41, 61–103
  arranging variables in experiment, 109–116
  assumption approach to independent variables, 116
  blocking independent variables, 115
  classifying treatment variables, 116–118
  confounding, 92–95

Validity/validity checking (*Cont.*):
  covariates (independent variables), 108
  crossed analytical models, 111–112
  dependent (response), 40, 62, 76, 105–108
  distributions, 116–118
  fixed vs. random effect, 117–118
  hierarchical design, arranging variables, 112–114
  incorporating independent variables in experiment, 110–112
  independent variables, 41, 105–118
  known, nonmanipulable nuisance variables, 116
  levels for treatment variables, 118–120
  limitation/control of independent variables, 114–115
  nesting independent variables, 112–114
  nuisance variables (independent variables), 108
  qualitative vs. quantitative variables, 117
  treatment variables (independent variables), 108
Variance, 79–83, 191, 194, 197–198, 234–240

Yields, 40
Youden square design, 102, 120, 221

ABOUT THE AUTHORS

JEFFREY T. LUFTIG is president of Luftig & Warren International in Southfield, Michigan, which develops and delivers training and consulting services in the Quality Sciences to businesses and industries throughout the U.S., Europe, Asia, South America, and Australia. Clients have included Alcoa, BF Goodrich, Hughes Aircraft, IBM, Ford Motor Company, and Motorola. A prolific author and lecturer, Dr. Luftig is the former Associate Dean of the College of Technology and Director of the Technology Services Center at Eastern Michigan University. He is a past recipient of the Endowed Chair for Research, Development, and Training in the Quality Sciences, funded by Ford Motor Company's Electrical and Electronics Division.

VICTORIA S. JORDAN, a Certified Quality Engineer, is a senior staff associate at Luftig & Warren International, Inc., where she assists service and manufacturing companies in implementing Total Quality, including strategic planning and policy deployment. She also provides management and consulting training to Fortune 500 companies implementing TQM. A former quality assurance manager at General Electric, Mrs. Jordan is a senior member and past section chairperson of the American Society for Quality Control and past chairperson of the program committee of the Nashville Quality Forum.

RECEIVED
ALUMNI LIBRARY
AUG 04 1998
Wentworth Institute of Technology
550 Huntington Avenue
Boston, Ma 02115-5998

Date Due

VC 734782		
VC 777290		

BRODART, CO.   Cat. No. 23-233-003   Printed in U.S.A.